CW00524436

ANCIENT RIVER ⎯ ⎯DING

THROUGH INDIA TO THE SOURCE OF THE GANGES

Published by

Librario Publishing Ltd

ISBN: 1-904440-25-8

Copies can be ordered via the Internet
www.librario.com

or from:

Brough House, Milton Brodie, Kinloss
Moray IV36 2UA
Tel /Fax No 01343 850 617

Printed and bound by Antony Rowe Ltd, Eastbourne

© 2003 Charles Clark
This book or any part of it may not be
reproduced without permission from the author

ANCIENT RIVER BENDING

THROUGH INDIA TO THE SOURCE OF THE GANGES

Charles V. Clark

Birth and death are doors through which you pass from one dream to another. All you are doing is going back and forth between this gross dream world and the finer astral dream world; between these two chambers of dream nightmares and dream pleasures.

Paramahansa Yogananda

1893 - 1952

Librario

Acknowledgements

I am grateful to Heather for her rapid and professional retyping of my much-altered manuscript. Thanks also to various friends for their encouragement. A passing nod, too, to my early reading, for leading me astray.

To those whose paths converged all-too briefly with mine along the way, wanderers mainly, following their own star, and whom I will probably not see again; without them this book would not have been written. <u>Shanti</u>.

CONTENTS

All photographs by the author, except the one marked by * which is by K. Ackermann.

GLOSSARY

Ahimsa	Harmlessness
Asana	Yoga posture, pose
Ashram	A spiritual centre
Asteya	Non-stealing
Avatar	One who attains union with Spirit and then returns to Earth
Avidya	Ignorance (of Divine Spirit)
Baba	Respectful form of address
Bhagavad Gita	An ancient Indian scripture
Brahman (Brahmin)	Absolute Spirit; Supreme Reality
Bhakti Yoga	Spiritual path based on devotion
Brahmacharya	Abstinence; one who lives in higher consciousness
Chakra	Psychic centre or wheel
Charpoy	String bed
Chela	Hindi word for 'disciple'
Chulha	Earth fireplace
Dharma	One's inherent duty
Dukkha	Buddhist term for suffering, unsatisfactoriness
Fakir	Magician
Ghat	Riverside cremation place
Guru	Spiritual Master
Hatha Yoga	Bodily purification practices
Karma	Effects of past actions
Karma Yoga	Selfless action
Kriya Yoga	An ancient spiritual science
Kundalini	Latent energy of a person
Laya Yoga	Union through perception of astral sounds
Lingam	Oval-shaped stone representing Siva
Mala	Bead rosary for counting
Mantra	Mystical root-word sounds
Maya	Transitional world of Nature
Mela	A fair, Khumb Mela
Moksha	Liberation from cycles of births and deaths
Nirvana	Enlightenment, Samadhi
Om (Aum)	The original sound (mantra)

Pandit	One learned in spiritual matters; commentator
Pani	Water
Parikrama	Circumambulation of holy mountain
Prana	Vital energy; bioenergy
Pranam	A form of greeting in India
Puja	Hindu worship
Raja Yoga	The "Royal" or highest path of Yoga based on Kriya Yoga
Rishis	Ancient seers
Sadhu	Holy man
Sangham	Holy place
Satsang	Spiritual instruction
Siddhis	Powers of a higher realm
Siva (Shiva)	Hindu god, "the Destroyer"
Sri	A title of respect
Swami	One who has renounced mundane experiences as a goal in life
Upanishads	Commentaries on the Vedas
Vedas	Oldest known religious texts, written about 5000 B.C
Vishnu	Hindu god; preserver of the universe
Yoga	Union of the individual soul with Spirit
Yogi	One who practises Yoga

TIBET

Yamnotri
Gangotri
Gaumukh
Barkot
Uttarkoshi
Bhagirathi
Kedarnath
Badrinath
Gaurikund
Joshimath
Mussoorie
Gawana
Tehri
Dehra Dun
Devaprayag
Srinagar
R. Alaknanda
Rishikesh
R. Ganges
GARHWAL
HIMALAYAS
Haridwar
Lansdowne
R. Ganges
DELHI
INDIA
Corbett
nat. Park
KUMAON
Almora
Ramnagar
The Garhwal Region
of Uttar Pradesh.

N

0   20 miles

0   20 Kms.   Haldwani

Chapter 1

Surfacing

As the light grip on my forearm slackened, the black-bearded Sikh flashed a disarming smile, to reveal the whitest of teeth: "Something has awakened within you – that is why you have come to my country." The cries of the street vendors and the bedlam of Delhi's Connaught Place faded into the background as the stranger's eyes fastened upon mine.

He inspected my face thoughtfully for a second, before continuing. "You very independent man…You been India already and you will come again." His alert gaze never wavered. "There is no need you search for a guru. You will meet one, a slightly balding man not so old. Where and when this will happen I cannot tell you."

With that, as a bead of sweat trickled down my cheek, he raised one arm in farewell salute. "Goodbye. You do not know me" – his eyes held my attention with mesmeric intensity – "I – know – you -."

In a trice he was gone, swallowed by the crowd. For a split second I caught a glimpse of a white turban receding, before the throng engulfed him for the final time. I looked about me. Mysterious to say the least.

It was a weird beginning, a surreal taste of much of what followed later. Our brief encounter set the rhythm, the pattern of future experiences: sublime moments closely juxtaposed with the ordinary; mundane days succeeded by the memorable. So is life. I tried to be receptive to whatever circumstances the cards dealt me. I looked for knowledge in a blade of grass or the glance of a passer-by, as well as in the utterances of reclusive mountain hermits. I found it pre-eminently in silence. On nearly every page I describe happenings and events in terms of "I did this", or "I that." And yet on numerous occasions I felt other than this physical body, these fallible, limited senses. Something told me I was living a kind of dream, from which at any moment I would awaken. Or was someone dreaming me?

As though at a given signal, the raucous commotion about me began to impress itself upon my consciousness even more forcibly than before as I, too, resumed my own journey.

I made off around the colonnaded outer circle of commercial buildings, erected in 1931 and named to commemorate the visit of the Duke of Connaught, brother of King George V. All life's variety, both of people and vehicles, flowed past me in a never-ending stream, and it was some time before I could cross over in order to reach the broad avenue of Chelmsford Road, one of several grand thoroughfares which radiate from the shopping and business centre like spokes from the hub of some giant wheel.

Market Road, when I reached it, was a blue-heat haze, a shimmering ravine of clogged pedal and motor rickshaws, a nose-twitching cocktail of rotting vegetables, exhaust gases and aromatic spices. A raw-boned sacred cow wandered by with enviable indifference, to scavenge amongst the wasp-striped *Vikram* taxis. A beggarman with one leg ending just below the knee squatted on the pavement, a hand stretched out, ravaged by leprosy.

I trudged on through the quivering air past stalls piled high with all manner of fruits or strewn with cheap cotton goods. On either side of the stuffy lane ran an irregular line of telegraph poles, some leaning at odd angles, and from which wires crossed and recrossed space in every direction. The bazaar seethed with many different nationalities of all ages and skin colours; stray dogs and suspiciously healthy-looking flies added to the restless hubbub of activity. I spotted a discarded carton racing as if on wheels along a shallow culvet. Only as it tipped down a foul-looking hole did I realise that the 'wheels' were actually the legs of a large rat.

At length and not before time I spotted an illuminated sign which I had been told to look out for: *HOTEL VIVEK.* My senses were still reeling from the assault of malodorous street fragrances, and it was a while before I dozed off in my ninety-rupee room on the second floor. Events of recent hours, some of a bizarre nature, drifted through my spinning mind.

On the plane from Europe I sat next to a retired Indian army colonel who told me he was on his way home after visiting relatives in America. I wasn't keen to promote my British origins while travelling in India – not all Indians have fond memories of the Raj – but curiosity got the better of me and I asked him whether we had left anything of lasting value after 1947.

He thought for a moment before responding.

"One. The British gave us independence without a battle for it. Two. They gave us the beginnings of a school system …opened colleges. There was very little of that before when most of the country was illiterate. We have built on that."

I didn't mention bureaucracy.

At the rear of the jumbo-jet squatted a bare-footed American Hari Krishna devotee, who informed me he was on an annual pilgrimage to his temple in Delhi. I quizzed him about his commitment to the spiritual path.

"The search for Truth is nothing whimsical," he replied. "You have to give up things, even friends and relatives. Sometimes they say things when you're making your commitment to the spiritual life; they can make things really difficult for you when they don't respect what you're doing. Even when we're living in a very pure way, close to God – at One – there comes a time when we have to make relationships with people. You can't just live according to some abstract idea of spiritual purity."

He had tried other religions and had found Buddhism "a bit cold." He spoke without obvious dogma. That is, until he began to offer advice on how to survive in India: "Watch the water here, it can kill you!" he warned me roundly. " – An' don't hit anyone! You take a rickshaw. Sometimes they ask ten rupees at the start, then want another five when you get where you're going! It's not worth arguing over it…He knows people and could make things really difficult for you. Hit the guy because you think he's ripping you off, an' there could be twenty of 'em piling on top of you."

"What about food?"

His eyes stared. "I eat only in the temples – never in the streets. After six or seven weeks the traffic an' pollution gets to me and I head back to the States for my health."

My rest refreshed me and two hours later I returned to the over-inhaled street in a more positive frame of mind. Now I was conscious of a further ingredient in the atmosphere besides questionable perfumes: a vibrancy, a pulsebeat of energy which I have rarely encountered in Western cities except, perhaps, during a festival.

Where to eat? Across the lane a restaurant's outside tables crowded with sprawling, footloose Western travellers, was the obvious answer. Seizing my opportunity when a gap appeared in the onrushing flood of rickshaws I hurried across and sat down at a tilted table close to the kerb and within inches of the passing traffic. On the edge of the gathering an old man, so thin I could almost see through him, scratched a haunting tune from a stringed assemblage made from recycled wood, metal and a length of wire. The travellers sat and chewed with a collective air of studied nonchalance.

The waiter appeared and I wrote down my order in English on the grubby pad. I glanced at my watch as I handed the pad back. One p.m. Many days stretched before me – time enough to wean myself off Western –style cuisine.

The restaurant was typical of travellers' meeting places the world over, a useful spot for a first-timer to exchange information on the usual subjects.

Eventually, a pony-tailed Australian wearing a bead necklace and mottle-hued waistcoat cocked his scraggy beard in my direction: "Bin here long, mate?"

I shook my head as my cornflakes arrived in a cracked bowl.

"About four hours."

The Aussie's eyes narrowed. "'sfunny, I'm sure I know you from some place...Rajasthan, Pushka?"

"My first day," I said, spooning up my cereal. I was heading to Almora in the morning, I added.

The desert town of Pushka was "a peaceful place", someone commented.

"Used to be you mean," another Aussie, with a fading yin-yang tattoo high on his sun-weathered shoulder, chipped in with the meaningful sagacity of the traveller who has been everywhere: cooled-out. "When you hear a place is peaceful in India," he continued, "it means it isn't. What they mean is, it used to be peaceful. Ten years ago when I went there, Pushka was peaceful...hardly any Westerners. Now everyone's started to go an' the result is it's no longer peaceful."

"Maybe I'll take a look, anyway," I replied. If I was centred in myself, then the best place was wherever I happened to be at the time. That wasn't the way it was every day, as you'll hear, but it's called living in the present.

It can be all too easy for Westerners to romanticize India, to imagine it as a place of perpetual spiritual sunshine – I myself am open to the charge from time to time. But a casual glance at a discarded Hindustani Times reveals a staggering statistic of suffering. During the previous two years nearly a thousand people had been killed on the railway tracks in and around the old capital of Delhi. Ninety per cent of those who died were cremated without being identified, the police issuing a routine 'public co-operation notice' but usually taking no further action.

There is no need to go in search of Eastern novelty in India: when you sit down in a public place it comes to you unbidden. My lop-sided table was a ringside seat for the passing show and I did not have long to wait for the first act.

A lone and lonely drum beat soon announced the imminent arrival of some solemn event. In a short while a funeral procession trod in slow, respectful silence past my table, the shrouded corpse held gently aloft by willing hands. I listened intently to the receding tap of the drum until at last its timeless note slipped beneath the surface sounds of the bazaar. Shortly, I decided to look for a less public restaurant.

Fifty yards down the lane I turned right past a cinema and slipped into a little icecream parlour-café whose walls were lined with spotless mirrors and garish posters depicting a variety of Hindu deities. Pleased to find it empty, I asked for a tea and took a seat to the rear, where an electric fan blew a steady breeze across the back of my neck and head.

It wasn't long before I had company. Looking up from my book, I saw a turban-clad, bearded man peering hesitantly over the glass-topped counter. Carrying a staff, sun-darkened and dressed in a heavy, homespun robe despite the heat, he looked like a prophet just descended from the sacred mountain. Catching my glance, he shuffled barefooted between the vacant tables towards me. I feared the worst when he abruptly slumped down at a nearby table. I was surprised to discover that he was not as Oriental as his appearance indicated for, as he turned his gaze to meet mine, I saw that his eyes were grey-blue, with a yellowish blur of fever.

"I don't suppose you'd mind buying me a drink? I've no money," he explained wearily in English, perhaps sensing a receptive ear. He was as English as I was and, for better or worse, I bought him a chai. The drink relaxed him somewhat and he began to unfold a tale of misadventure caused, so I understood, through no fault of his:

"I arrived in Delhi a month ago. I had no money since all my belongings, including my passport were stolen from my hotel room in Varanasi. I think the hotel manager was involved but I couldn't prove anything. The police showed little interest."

Varanasi is India's holiest city, but he was forced to wander the streets during the daytime, scrimping money from passers-by and sleeping rough in the railway station at night. Money had failed to arrive from home, his visa was about to expire, and he was fearful of being thrown into gaol if he was in the country after its expiry date.

The oppressive Delhi heat was beginning to get him down, he said:

"When you don't have a hotel and money, officialdom treats you as a near-vagrant. I have to keep in the shade. Then the police move me on." His brow furrowed, making him look much older than his mid-thirties. "I'm thinking of jumping the 10 o'clock night train to Rishikesh."

When I raised an eyebrow, he added, "You just pile into one of the unreserved compartments and stay there the rest of the journey...I need to get away from this heat for a few days." He spoke briefly about the Himalayas and the inspiration to be found amongst the peaks and valleys of the world's highest mountain range. I reached for my money belt, but he stayed me with a rueful look:

"Thanks, but I've done all I can. It's just a question of waiting."

Draining his glass he rose to his feet, extending a bony hand as he did so. "Thanks again – most Westerners just walk away when I approach them."

There was no self-pity in his voice.

I thought he would decline, but I shoved a ten-rupee note into his hand. Instead, he folded it and tucked it away inside his robes, saying he would use it later to buy curds. Just as he reached the exit I called after him: "Don't give in!" Before vanishing, he half-turned and shot me a wan smile and final words, "Frank...London."

I looked out for him in the streets later, but I guess he was true to his word and had jumped the train.

Next morning I took a pedal rickshaw to the inter-state bus station close by Kashmiri Gate. Despite assurances to the contrary at Hotel Vivek, the ticket office provided me with an abrupt awakening – there was no direct bus to Almora, the former Raj hill station in the Himalayan foothills to the north east of Delhi. I would have to head for the nearer township of Ramnagar first, the ticket clerk informed me.

Two hours to wait. It was six a.m. and, still groggy from my early rising I dropped onto a chair in the grubby station café-kiosk. The bus was half an hour late, but I clambered aboard grateful at least to be on my way.

By the time the bus rattled into the long main street of Ramnagar, a dusty, bustling market town on the edge of the Himalayas in Uttar Pradesh, my lips were salty and the back of my cotton T-shirt clung to me like a second skin. I had seen little enough of the countryside through the smudgy windows, yet a vivid memory lingered of an endless nullity of landscape; of livings scratched out in the parched fields, and of earth which seemed to have been oven-baked since the dawn of time. The monsoon was months away.

I retrieved my rucksack from the roof and set off up the street. Fifty yards away a welcome signboard caught my attention.

I stumbled towards it, eager to escape the cacophony of market day, not to mention the energy-draining heat. Inside I found shade but the noise level was scarcely lower than without: every table, so it appeared, was loud with the tumultuous babble of heated conversations, conducted in a language of which I as yet had little acquaintance. I dropped behind a vacant table near the restaurant door, jamming my rucksack behind me against the wall. Overhead a pair of fans languidly wafted the humid air about, but with minimal effect. At the far end of the room, a harassed balding man who proved to be the manager, was remonstrating with someone within the kitchen. By his shoulder, a wasted Western traveller with short peroxide hair and a grubby sleeveless vest, kept repeating, "Two rupees man, you owe me two rupees," over and over again like a jammed tape. If destiny kept such places in order to test the self-possession and progress of those who claimed allegiance to the yogic path, then I had certainly come to the right door.

With my pale skin announcing my recent arrival from the cooler regions of the northern hemisphere, I could but wonder what uncommon urge had caused me to part with a significant amount of money in exchange for the dubious benefits of Hotel Govind. At that moment, however, a cotton shoulder bag clouted my shoulder and another foreigner dumped himself unceremoniously in the seat opposite. Colourless eyes appraised me for a second, before a square hand reached across the narrow table towards me. "Nixon, Richard M. Nixon."

His clothing was nondescript wanderer.

"Vancouver, B.C.," he announced abruptly. "Been travelling round India for seven months now, an' Jesus – they're alus demanding to know my goddamn name! I stopped telling 'em – that's why I call myself Nixon. Why should they invade my private space just when they want?"

I conjectured what his passport name might be but didn't ask.

"Dharamsala," he ejected the word suddenly as if fired from a gun " – I heard that's a cool place to hang out, yeah?"

I had been to the lofty hill station once in 1991 and I told him of its wonderful location on the forested flanks of 16,000 foot mountains. A wary expression crossed his face when I mentioned the Tibetan health clinic.

A local teenager eavesdropping on our exchange at the next table leant across cheerfully: "You are Richard Nixon?"

The Canadian looked ready to vaporise and turned a thunderous look upon his inquisitor. "That's right – just join in without asking, will you!?" His voice was heavy with measured sarcasm.

The youth remained undeterred: "Your address, please? I write to you – you are my friend."

Nixon sprung wildly to his feet, flailing the sweat-charged air with one arm. "This country !" he yelled, " I ordered an hour ago…It's the goddamn manager who stops this place from functioning properly!"

I stayed about two hours more in the restaurant, fielding Nixon's sporadic outbursts with as much humour as I could muster, while wondering if I would become as hysterical after six months in India. Our orders duly appeared following another protracted interval. The Canadian gulped his down, complained bitterly about "the worst meal of the century", and disappeared to his room.

Next morning I discovered I had missed the only bus to Almora and would have to make the best of an unintended day in Ramnagar. Once I overcame my initial annoyance at missing the bus, however, I decided it would be best to treat the situation as an opportunity rather than a day lost.

Despite Nixon's withering character assessment of the entire Indian nation, I found the manager of Hotel Govind to be a helpful, if highly anxious, man. If a hundred foreigners should visit Ramnagar, he told me, eighty would be sure to stay at his hotel. I had not noticed any other accommodation signboards on the street, but the polite side of my nature stopped me from enquiring as to the extent of the competition. The building has been erected by an uncle, a district contractor in 1947. Although at present it had only four guest rooms, he had ambitious plans to build a far larger hotel within the next two years; it would have thirty to forty rooms, car parking, a market on the ground floor, plus an excursion vehicle for visitors.

He offered to take me to see his plantation on the back of his motor bike.

My lofty expectations of Indians received a rude awakening when I was introduced to a confident and much younger brother. He breezily put me right about the central crisis of modern India. "Religion is from the past," he informed me. "It is only for old people, the mantelpiece." He had studied technology at university for six years and knew that the popular, pot-bellied deity, Ganeshi, would have little or nothing to do with India's future. Unlike his brother, he wore Western dress and sounded more from that part of the world than I did. Around mid morning I booked a second night at the hotel and informed the manager that I was going to explore the nearby countryside for a few hours. He warned me to be careful in the forest as it contained many wild animals, including tiger.

The narrow lanes across the road from the hotel soon led me to the decaying edge of town where nervy, coarse-haired dogs yelped amongst low heaps of rotting vegetable matter. I crossed an arid strip of dark, undulating waste-land to the concrete embankment of a canal containing two or three feet of sluggish water. I set a strolling pace, partly due to the steadily rising temperature, but also because I had a friction sore caused by the rubber thong of my flip-flops. In my bag I carried my wineskin water bottle, some fresh fruit and the compressed remains of a piece of cheese. I headed left towards a ragged canopy of trees which rose above the distant outline of a metal girder bridge. I soon came to a dilapidated fence enclosure guarding a seated statue of the Moon-god of the mountains, God of the gods, Siva. Resplendent in a glorious coating of cosmic blue, the deity seemed to glow with a surreal intensity as though illumined from within.

The colour blue, as well as signifying the universe over which Hindus believe Siva to reign as Lord and Protector, also symbolizes the occasion when He swallowed the poison of Vasuki, King of the *Nagas* or serpents. According to legend, had he not held the poison in his throat lest He himself be destroyed, the whole of humanity would have perished.

I noted the symbolic paraphernalia which invariably form part of such shrines: the king cobra winding about his head and shoulders denoting His mastery over the human passion anger; the raised right hand offering reassurance, while the drum in His other hand denotes that God is the source of sound, *Nada-Brahman*. The tiger skin upon which Siva sits demonstrates His conquest of aggression and greed. Sometimes He is represented with four arms as 'Lord of the Dance', Siva Nataraja. The four arms indicate He holds life and death, good and evil, in balance. The dancing is the energy of the universe which brings destruction, but only so that it may be created again and again through the eternal cycle of Time. Squirming beneath His dancing feet was a grimacing dwarf, representing human ignorance. The torch in the left hand is a reminder of the inevitable destruction of all living things, while encircling flames represent the eternal cycle of Time which has no beginning and no end.

The custom of sometimes showing Siva as carrying weapons, the bow and arrow, trident or javelin recalls His place in the ancient myths as Pasupati, 'Lord of the Animals', as well as creator and destroyer of the universe. Hunters of long ago used magical powers, *Tantra*, to protect themselves from danger when they ventured into gloomy forests in search of prey. Prayers to the gods of the animal world would hopefully increase their chances of success. Those secret arts survive to this day as a means of acquiring transcendental powers and to achieve *moksha*, or salvation. The trident was probably an early hunting weapon for catching fish.

In the temples Siva is most prominently worshipped in the form of a *Lingam*. A short pillar usually regarded as the male organ, though some prefer to see its ambiguity as symbolic of the God's ability to assume all shapes. Often it stands in a shallow basin, the *yoni* or female organ of generation, their union being intended to represent the divine consummation with the Life Source. As Lord of the Eternal Snows, Siva is also God of the yogis, those beings who seek out quiet places amongst the hills and mountains to practise the physical and spiritual austerities of the Yogic path.

Followers of Siva, who usually paint three horizontal stripes either side of the psychic 'third eye' do not worship Him out of a love of destruction ('Siva the destroyer'), but out of an understanding that all things will come to pass, and from that passing will come a new beginning. The deity is the reconciliation of all opposites, creation and destruction, male and female, good and evil.

For Hindus it has never been a problem to have so many gods, or that some of them can assume a bewildering range of guises. Each god indicates a path, a line of stepping stones to a final understanding of the One Great Being. The fact that Siva appears to us in so many forms merely indicates the illusory and temporal nature of life. His different faces show us the various aspects of His 'character' in which we may also see ourselves. Siva presents Himself as a fledgling youth in the form of *Dakshinamurthi*, in order to show that the soul is not bound by the phenomenal world of age and aging. Reality does not grow a white beard.

Perhaps more than any other spiritual path Hinduism resists definition. It has no founder like Christ or Gotama Buddha who both proclaimed a creed. Nor is there a firm date when it can be said to have started. The earliest Hindu scriptures can be traced back to around 1500 BC, though the teachings themselves must have been handed down for many centuries before that. Its assimilation of pre-historical tribal religions gave it an astonishing assortment of deities, customs and rituals, almost beyond the Westerner's powers to penetrate. 'Popular' Hinduism to be found in the villages still retains numerous attitudes from long ago, such as a local goddess worshipped by childless women. To cater for the great changes taking place in Indian society new, splendidly bountiful gods are available to meet every conceivable requirement. A depressed student, for instance, may appeal to the deity for examination success.

Broadly, there is a significant difference between the Hinduism of the artless villager and that of the more educated person familiar with the sacred texts. But though there is no formal creed as such, a Hindu is socially and ritually bound to the group into which he or she is born, the familiar and now much-attacked, social caste system.

Western religious tradition has been a prophetic one emphasising God's revelation to Man, whereas Hinduism has followed a mystical way which lays emphasis more upon the discovery of God from within. The Western scepticism towards the reality of the inner path surely arises from a rooted attachment to an overriding belief in the reality of the material world?

I was to listen to much more on this interesting subject at Rishikesh, after my visit to Almora was over.

I lingered close by Siva's companiable presence for several minutes. By chance, a mass of pure white clouds ballooned over the nearby forested hills which seemed as if they simulated the far-distant Himalayan snow peaks, as the shining moon of wisdom hanging in the deity's hair denoted His asceticism and purity.

I moved on, padding slowly towards the shimmering parapet of the metal bridge. It was monotonous walking, but it felt good to be the only human being in sight.

Half an hour's walk brought me close to the bridge. It carried a main road from the town as well as functioning as a barrage across the river, its stark lines relieved at either end by a display of chest-high bushes ablaze with blooms of crimson and purple. I waited for a truck to career across before plodding over to the trees, where a well-beaten track looked as if it might lead somewhere.

The trail took me along the forest edge, winding through scattered groves of lifeless-looking bush, while above towered a variety of spindly trees whose names I did not know. Some had branches ending in spectacular sprays of scarlet flowerets. For the first time I became conscious of the continuous twittering of birds both around me and overhead. One, with a turquoise body and black wing-tips being particularly memorable.

Before long I heard voices and soon a line of sari-clad village women approached. One made a jaw of her fingers and warned me of "animals", snapping her hand towards the thickest part of the forest as she spoke. About this point the trail swung to the left and I soon found myself close to the river once more. Here I heard intermittent bursts of children's laughter ringing amongst the boulders.

The river was no more than a stream, it being the dry season in the northern plains. With their coppery skin glistening with water, two young boys ran naked amongst the shallow pools, shrieking, and darting nimbly from one placer to another. If they saw me at all they took no notice and after a few moments I sat down for a short halt amongst the polished boulders of the river-bed. A faint but cooling breeze fanned my cheeks, while the stream's lively note seemed to pluck a corresponding chord in the instrument of my mind, and I began to feel a deeper rapport with my surroundings.

Hardly had the sweat begun to dry on my brow when a loud rush of wings shot directly overhead and wheeled upstream, as at least fifty parrots disappeared from view. I took a pull from my wineskin before setting off once more, crossing the trail to plunge directly into the forest. There was little or no dead timber or undergrowth to impede my passage; most of it seemed to have been cleared for firewood. At first, visibility was twenty or thirty yards in each direction, but gradually the trees crowded closer as the evidence of wood-gathering diminished. The breeze was no more and after thirty minutes progress the urge to sit down again returned.

I had no idea of the passage to time when I woke from a light doze, but suddenly – for no apparent reason – I was very wide awake indeed. Something had jerked me into instant wakefulness – but what?

I glanced about me into the light-dappled foliage, looking and listening hard. The two warnings flashed unnervingly through my mind. Someone or something had drawn close without my knowing it.

An interminable minute trickled by. Still I could detect no movement. Was my imagination playing tricks? When I exhaled the sound was startling against the impassive silence which now pressed in even closer. A rivulet of sweat lengthened under my armpit as fear, the uncertainty of not knowing, welled-up inside me. I tried not to dwell on the fact that I was alone in Corbett country. I could detect nothing amongst those shadows and filtered light; humps and broken branches assumed all manner of menacing shapes. I couldn't see anything – but I sensed something.

I recovered my bag and rose cautiously to my feet. Suddenly – almost directly overhead – there was a whisper of movement! For a split second a pair of unblinking eyes stared into mine – then they had gone. I caught a glimpse of a grey shoulder, rounded yet powerful, then it, too, disappeared. I froze as a harsh chattering accompanied by a crashing of branches melted rapidly away, as whatever it was made off deeper into the trees. The sounds grew fainter and fainter as my racing pulsebeat slowly returned to normal.

At the restaurant I was disappointed to learn that the irascible Nixon had left town soon after noon on a 350 Royal Enfield, vowing never to return to "the Pits" whose shortcomings he hadn't experienced "since Morocco".

In my absence a few more travellers had drifted in. A German woman asked if I always went into nature alone. For a moment I thought to respond by citing the American naturalist John Muir's creed about experiencing the wilderness on one's own: "All other travel is mere dust, hotels, baggage and chatter", but in the circumstances it sounded pretentious and I replied with an evasive, "usually". Anyway, a niggle had arisen in my mind about Muir. I heard he didn't have much time for the <u>humans</u> who lived in the wilderness, the Indians. It takes all types.

Next morning I left the hotel whilst most of the foreigners were still in bed and made my way down to the bus stand. I was an hour early but Almora was obviously a popular destination with the locals, since most of the better seats were already taken. The good news was that the ticket for the eight-hour journey cost me scarcely more than a pound. I squeezed into an empty seat away from the wheel arches and slid back the greasy window.

I had clued-up a fact or two about my destination. Once much frequented by the occupying British when escape from the torrid plains became imperative, it was, the guidebook promised, "a finely situated hill-station at an altitude of 6000 feet". Until the Ghurka war of 1815 it had been part of the Kingdom of Nepal. I had no great expectations, but hoped both for impressive views of the snow-clad Himalayan giants further north and to remove myself from such nocturnal crawlies which have a liking for human footwear in warmer, lower districts.

The region was also said to be favoured by a number of hut-dwelling yogis and others who wandered from shrine to shrine and from temple to temple on *yatra* or pilgrimage. If our paths should cross they might well be experiences of some value. I had heard a number of mysterious accounts concerning some of these ascetic solitaries and was interested to know more. However, I didn't want to fall into the trap of 'holy men chasing' which ensnares not a few Westerners, who roam hither and thither across the dusty sub-continent on their endless quests.

Might is right on Indian roads and pedestrians, lone cows, scooters and bullock carts, all eased to one side as the bus ploughed on regardless of hazard warning signs and right-of-way. My fiercely –upright seat had surely been designed with extreme discomfort in mind, but I eased backwards, telling myself that the eight-hour journey would soon pass. A few minutes beyond the metal bridge I had crossed the previous day, an ominous rumble came from under my feet and the bus jerked to a halt on an inclined hairpin. During the enforced halt, several more people materialised out of the bordering forest and entered the bus, squatting amongst the luggage piled in the narrow aisle. The door flapped open at every bend, which I suppose was a novel way of reducing the interior temperature by a merciful few degrees.

We were, I suddenly remembered, passing through the district of Kumaon, made famous by the legendary Jim Corbett, whose book, *Maneaters of Kumaon*, I had read avidly whilst a teenager at school. The road was fringed by tangled woods, punctuated occasionally by gaps through which I could see patches of terraced cultivation and more forested tiger country. Today, as elsewhere in the world, this magnificent creature has been hunted almost to the point of no return. Hopefully, the vigorous moves to protect this and other species have not come too late.

The hours, the bends and the occasional small town or village slipped by in a haze of encircling views and grinding of low gears. Once there was a sudden craning of necks and excitable babble from the off-side seats. Half-rising to my feet, I caught a glimpse of the cause of so much animation: a light truck had plunged off the road and hung suspended amongst branches 100 feet down the near vertical slope. I ate a light meal at a stall on the main street of a ramshackle village surrounded by tiers of terraced rice fields, then it was on and up to the next place, raking up passengers and packing them in: the bus was rarely less than over-full. Once or twice I caught tantalizing glimpses of higher hills with vaporous valleys hanging between.

As we gained altitude, the temperature gradually dropped, and after some three hours, we found ourselves proceeding through an opaque, watery mist which obscured the windows and created fantastical, shrouded forms in the increasingly dank forest. I wondered idly how the driver managed to negotiate such a tortuous road without windscreen wipers whilst continuing a non-stop conversation with his neighbours in the adjoining seats. Perhaps our safe passage was being guarded by the little effigy of Ganeshi, the elephant-headed god and 'remover of obstacles', who perched with serene detachment on his dashboard. I slipped a pullover on top of my T-shirt and pushed the frayed edge of the road from my mind.

At last I spotted a several-stringed necklace of lights sparkling about the summit of a darkening hill on the far side of a mist-hung valley. "Almora?" I queried of the passenger next to me, a villager with a quick smile and a droopy moustache.

"Yes, yes!" he nodded helpfully, revealing teeth stained by a long habit of betal (*paan*) chewing.

"Which country are you coming from?" he continued, as if a dam had burst. "You are travelling alone? What is your occupation?"

Another hour crawled by before, slowly, the lights of Almora town began to suggest a more material world of streets and the dim forms of buildings. Finally, we rattled into an ancient lane, crunching to a halt much as we had departed – amongst a throng of faces.

I swiftly recovered my rucksack from the roof and two minutes later – more by accident than by instinct – found myself before a stone-built, three storey hotel not far from the heaving bus stand.

The manager flung open a door round the corner from reception on the ground floor.

I disliked it instantly, a room with all the virtues of a punishment cell or an immigrants' hostel: a rock hard bed, bare stone floor, 10 foot high whitewashed walls and a solitary lightbulb dangling under a cracked shade.

I shook my head. "Have you one with a window, a view?"

The manager eyed me dubiously for a second, as though asking for a room with a view was a character defect of a quirky foreigner. I followed him up to the first floor landing where he proceeded to wrestle with a second door. The room was much like the previous one, although some spendthrift had placed a table lamp by the bed.

"The window?" I asked, thinking it would be good to wake up to a view of the world's highest mountain range.

He gestured towards the ceiling with a wave of his hand.

"Window room, sir: plenty light. You sign book downstairs. Forty rupees."

In the empty foyer the daywatchman, dressed in a kind of stage soldier's uniform, snapped to attention as I approached and greeted me with an earnest salute. After this had happened two or three times, I began to think I must be a visiting celebrity and that a limousine with chauffeur-held door would be parked at the foot of the hotel steps, waiting to whisk me away to a swish restaurant.

No such luck. I trudged up the puddly, gleaming street in a thin drizzle and ate in solitary splendour at the Madras Café. Next morning I returned there for breakfast. The weather had deteriorated further and the blistering heat of Ramnagar already seemed a long time ago. It was dreary, English-type weather with lowering grey clouds blotting the mountains from view. But the night's rest had been a good tonic and I felt unaffected by the dispiriting conditions.

At the far end of the long, cloth-covered table sat another Westerner, the first I had seen since leaving Ramnagar. He gave me a friendly nod but initially remained silent, eating reflectively from the plate in front of him, and every now and again glancing into space with an open, clear-eyed look. In his mid-twenties, he wore shoulder-length hair tied back with a coloured band. The prayer beads about his neck, the faded *lungi* and the sun-darkened skin suggested a seasoned wanderer.

We shared the same table for only half an hour, but in the little he did tell me about himself, it was clear that, with the possible exception of Frank, he was very different from any of the Westerners I had met so far. He wasted no words in conversational pleasantries, nor was he interested in telling me which country he was from, or in knowing mine.

"I am just a man, a being of the universe," he told me.

He said he walked everywhere, sleeping in the woods and travelling from village to village, where he was invariably received with kindness and was given food. In the past he had tried working 'straight' in Canada and the States, but had given it up because such employment meant being something other than himself. He liked to be close to nature and told me of a place only an hour's walk away where I could obtain a simple room for only five rupees a night.

"People stay there a long time," he explained with what sounded like a faint French accent. I said I respected his commitment. The wanderer's eyes glowed.

"The task is...to get free," his eyes lifted for a moment to the passing scene, "from *Maya*. We are all on the same journey, whether we know it or not."

There was another reflective pause during which I sensed he had said all he wanted to say. Sure enough, a few moments later he rose easily to his feet to take his leave, his gaze turning towards the open door. He looked into my face, pressed both palms lightly together before his chest, saying, "I hope you have a good life."

Then he turned and walked out of the restaurant, a light bedroll folded over one shoulder, a sadhu's brass water pot in one hand and a staff in the other.

The term 'sadhu' embraces a whole range of religious adepts, from the practitioners of mysterious occult powers, such as magicians, sorcerers and contortionists, to the genuine holy men owning little more than a water pot and a staff. The true ones dress in orange or saffron or a simple loin cloth. Others, such as the *digambaras* or the *nagas*, see themselves as clothed in air or space and go naked. Some smear themselves with ashes to signal that they are free from the fear of death. Genuine or not, they are a dwindling band, slowly succumbing, as did their distant brothers, the wandering, mendicant monks of medieval Europe, to the onward march of modernisation.

For the most part, Western visitors to India do not have the time for such a quest and regard sadhus, at best, with amused indifference. But it is easy to forget that what we too readily dismiss with our preconceptions, is part of a respected tradition which can be traced back some 5000 years, to pre-Vedic days. 'Modern' city Indians, too, are often suspicious of the sadhu and his calling. Not without good reason. Those who have simply tired of worldly occupations and find the 'holy' life of *baksheesh* and regular attendance at festivals a more congenial lifestyle, are seen as parasitic, making no economic contribution to the country. Some possess an eye for Westerners pursuing the spiritual path and take advantage of their vulnerability. Others attempt to demonstrate their dedication by holding an arm aloft for ten years with nails as long as an anteater's. That India can no longer support its swarms of holy men, is a widespread attitude.

Yet the corrosive effect of over-scepticism needs to be tempered with tolerance and understanding, lest all be tarred with the same brush. Some, though they are not always easy to find, have given up goods, worldly positions and property to devote the remaining years of their lives to a genuine search for God.

A Hindu business man waiting in a bus queue at Connaught Place explained.

"I am doubtful of most of them, though the spiritual search can have its reward. For every one sadhu who is true, there are a hundred who are not. God is big business in India." His tone became more serious. "If you want to find God in this lifetime you will need a guru to help you. But you will have to deserve one before you will meet him. If you are not sincere and have not already made some effort yourself, then you and he will not meet. There would be no point because you would not learn anything."

In the neighbouring Buddhist country of Thailand there is still great respect for mendicancy, the traditional monk's way of life in which he may only live on what is freely offered on his village rounds. The term 'tudong' is used to denote the journeyings of such Buddhists, whose wanderings may last weeks, months, or even years. They live by precepts which most of us would find daunting for a weekend, never mind a lifetime. Using their lives of quietude for reflection, they are able to offer wise counsel in return for alms. In England in recent years, this ancient practice has been revived by certain Theravadan *viharas* of monks and nuns on a limited scale, and with some success.

Was this spiritual path, like the Raj, merely part of the fantasy of the past, a bitter-sweet memory of what India in its 'development' has lost? I hoped to find on my travels some evidence that the ancient, mystical side of this bewitching country still counted, that it was more than a mere creation of flabby intellects content to turn yet again, the yellowing pages of nostalgia.

A few minutes later I followed the Western sadhu from the café and headed up the lane in the general direction of the town's highest point. The road levelled, then dipped past a couple of whitewashed shacks with corrugated metal roofs. Below to the left, the steep hillside was split by slanting ravines. It was a good lookout point and I stopped to take advantage.

Across the immediate valley a metalled road cut steeply down a hillside riven by a succession of eroded gullies; on either side of the point where it dropped out of view, the lie of a ridge was marked by a broken line of trees on one side and untouched forest on the other. Two parallel valleys crossed the middle distance, whilst beyond them, hung with ragged shreds of mist, was a further wave-like succession of valleys and forested ridges. The sky was a lowering grey and reminded me much of the continent I had recently left. I turned my head to the north, but there was no scenic vista of snow peaks: all was an impenetrable leaden monotone. If only the sun would show itself! Not too far distant also, was the famous Valley of Flowers, important for Sikhs as well as for the phenomenal numbers of plants which blossom there during July and August.

At length my gaze returned to rest on a deodar pine, whose graceful limbs dripped with ragged beards of lichen. According to legend, the deodar is the most spiritual of trees and the favourite of the gods. A few feet back from the edge of a cliff it stood like some ancient monarch surveying his kingdom. It seemed to breathe, to emanate some hidden understanding which reached into me like a cool hand. Slowly I began to feel connected to the tree, together with the little heap of angular pebbles which someone had placed at its foot. The feeling of being divided by contrary desires began to melt away as some inexpressible stillness took its place. I could not help but wonder again, if the wandering sadhu's path and my own had crossed for some purpose. I felt immediately happier and, after a few moments more, I turned about and began to head back. Tomorrow I would leave Almora.

The town is actually one of the staging posts en route for Mt Kailas in Western Tibet. Reached only after a gruelling journey of some fifteen hundred miles or so, the mountain is probably the most holy place for Hindus and Tibetan Buddhists alike in the whole of Asia. The former regard the peak as the retreat of Lord Siva and his consort Parvati. To Buddhists, Kailas is known as *Gang Rinpoche* (Precious Snow Peak) or simply as *Ti-se*, 'The Pinnacle'. It is the Mt Meru, the lofty seat for countless deities and place of supreme wisdom and cosmic power. The struggle to get there stands for a kind of wearing down of the ego; the pilgrim's sense of separateness slowly dissolves. Many who return from such barren wastes remark how the intuitive faculties of the mind have been opened and energised by the strange powers which are claimed to reside amongst those seemingly hostile snows. Few Westerners had been there until comparatively recently, whilst Chinese obduracy has not greatly improved prospects for the future.

It was my ambition to make that goal one day, but in the meantime the ultimate destination of my present wanderings was waiting many miles to the north-west in the Indian Himalayas: 'The Cow's Mouth' or the source of the River Ganges, the sub-continent's most sacred river. I had tried to reach it early in 1991 but had to turn back in the face of deep snow and approaching storms. It had always been my intention to return for another attempt.

Wandering back the way I had come, I spotted a soldier cradling a heavy rifle beneath a sign, 'State Bank of India'. It was a good excuse to change money.

At my hotel the manager asked whether I had anything for sale. I shook my head, telling him I was leaving next morning and wanted to know the bus departure times to Rishikesh. His face split with genuine enthusiasm:

"Rishikesh very holy place – many Indian people go there." His eyes shifted back to my shoulder bag. "<u>Nothing</u> for sale, sir?"

"Only a watch," I said, suddenly remembering a cheap, Woolworth's digital, which an out of date friend had assured me would fetch a good price in India.

The manager chuckled good-naturedly and pulled back his shirt sleeve to show a knobbled, chromium-plated time piece. "One is enough, sir." He smiled broadly when I recommended mine for his other wrist. "It is a jacket I am most needing, sir."

I changed the subject: "This weather, how long will it last?"

He glanced at the streaming window with a shrug. "At this time of year, sir- who can tell?" He spread one hand before him: "One day, two days…two weeks. It is like English weather, sir." His frown vanished with a vision of better days ahead. "You come back in two months… three months – stay at my hotel, sir. Plenty sunshine when you come, then."

The daywatchman nodded indulgently as his employer pumped my hand as though I was a relative returned from the dead.

That evening I took a last stroll around the dusk-laden upper lanes of old Almora; the power of darkness had restored the decaying walls to something of their lost glory. Spirits seemed to lurk amongst those shadow-filled alleys. An odd mixture of feelings arose within me when I spotted an unmistakeable silhouette rising over the immediate horizon of flat-roofed shops and dwellings. It was an English village church. An aura of desolation seemed to emanate from it and I hurried on before its gloomy outline could stir too many memories.

A dog barked a warning from behind a door and was quieted by a woman's voice. Beneath the pale glow of a lantern a cluster of locals stood smoking and conversing in quiet undertones. A few yards further on the lane opened on to a broad mall and I discovered the reason for the scarcity of people in the streets.

Most of the mall's windows were shuttered for the night, but one shop was very obviously open.

Despite the chilliness of the damp evening, over a hundred people were pressed close to its illuminated front. The object of their attention in an otherwise deserted mall was – a colour TV. I approached through the semi-gloom and joined the edge of the assembly.

"What is it?" I asked of the person nearest to me, a man whose face was lit by a kind of ecstatic attention.

He beamed at a foreigner's interest: "It is *Mahábhárata*, famous Indian story; every week…" His gaze, like everyone else's returned to the screen.

This great Indian epic, or cosmic drama, is the longest poem ever written, containing well over two hundred thousand lines. It is said to have been composed around 200 BC by a *rishi* called Vyàsa, though whether there was only one vyàsa, or 'arranger' around at the time is open to speculation. The tales are of hero-gods who become men and of legends of the saints. Terrifying battles are fought between forces of good and evil, but so full is it of readily recognisable personalities and human encounters that it is very accessible to a wide range of audience sophistication. The heart of the *Mahábhárata*, the *Bhagavad Gitã* or 'Song of the Lord', shows in a profound way how human conflict, both social and individual, can be resolved and may be interpreted on a number of levels. The Gita explains the deep mysteries of life and death and teaches the truths spoken by the *rishis*, or sages, long ago.

A swift glance at the rapt, attentive faces about me told me that Krishna, the charioteer, Rama, Sita and all the rest, were far more than mere stage personifications from some fictional drama and ancient mythology; it was clear that the 'Song' reached to the very hearts of Hindus. The characters, living beings whose dilemmas, both human and divine, spoke to them in their everyday situations in a very real way.

The action takes place on the battlefield of the World and includes all the diversity of types of character that life provides. In the dialogue between Arjuna and Krishna, for example, nearly fifteen hundred lines of verse are concerned with the subject of salvation. Krishna, who is also the god Vishnu embodied in human form, advises the famous warrior Arjuna on the proper course of action he must follow to win salvation. Only if he seeks Him with all his strength will he be released from *Maya*, the human condition of doubt and perplexity, and find his true Self!

"Undoubtedly, O mighty-armed one (Arjuna), the mind is restless and hard to control; yet by practice and dispassion, O son of Kunti, it is controlled."

And later, "He who does the prescribed work without caring for its fruit, is a *Sannyasi* as also a Yogi, and not he who is without the (sacred) fire and without action".

Filled with doubts Arjuna (the soul), seeks Krishna's direction at every turn. His agonized predicament is a consequence of his own past actions and attachments to selfish desires and material things in this life. Thus in Hinduism the causes of suffering differ from the central creed of Christianity which says our dissatisfied state is a result of sin and disobedience against God. The story is no dusty tract of learned philosophy, but is filled with the breath of divine love; were it otherwise, the television would undoubtedly have been playing to an empty house. Yet even in the corner of that market place in the Himalayas I could sense how its spirit, like an electric current, embraced all those who stood there, filling them with an unshakeable feeling of community.

It is other in the West, where we tend to grasp at the treetops with our senses, while insisting they are the roots. A tiny part of me still felt an outsider and later on I wondered if Jung was correct to claim that we can never enter into the *communico spiritus* of a foreign culture? Was I no more than one of his "mediocre minds" whom he appears to dismiss as losing themselves "in the surface and externals of a foreign culture"?

After ten minutes camaraderie I slipped away into the anonymity of an unlit side lane towards my hotel. Rain once more began to bead gently on my outer clothing as I stumbled from one unlit, time-stained passage to another. It was much too early in my journey to start cudgelling my brain for answers to such questions. Would I one day splatter myself with ashes, who could tell?

My room wasn't the cosiest to return to at night. I lay on my bed with only the feeble glow of a street lamp for companionship, and wished I had brought some tapes with which to divert myself. Eventually I fell asleep.

I awoke once during the darkness thinking I was in a boat at sea. After a moment's confusion I realised that the beating sound in my ears was a high wind flinging rain against the invisible window pane. I drew the sleeping bag up to my chin and shut the storm from my mind.

Doubts faded with daybreak. I arose about seven – a quick glance outside confirming I had made the right decision to leave, the world beyond the immediate confines of the lane being lost in a sea of clouds and lashing rain. The deluge sprang from roofs and rusting fall pipes, spouted from drains and hammered upon glistening umbrellas. Rivulets ran from every crevice, cascaded together, and flowed on, an unstoppable sluice which under normal conditions was the street. I peered through the deluge towards the bus stand – and received a jolt - the bus was already loading up!

Despite the size of the throng, however, it was little more than half full when we pulled from the terminus like some lumbering scrapyard amphibian. Within ten minutes a slow, drip-drip of water began to fall from an ill-fitting window frame directly on to my upper trouser leg.

Chapter 2

Gateway To The Gods

Where the River Ganges spills out of the Himalayas on to the wide plains of northern India stands the old town of Haridwar, also known as Gangadwar. The bus had dropped me some distance from the town centre on a pot-holed and oil-darkened strip of earth euphemistically called the 'bus station'. On the opposite side of the road was the railway station. I had been assured that the journey from Ramnagar would take three hours at most, but a combination of a narrow, busy road frequently made narrower by repair works and dawdling bullock carts, bicycles and a variety of motorised vehicles, extended the time to nearer seven. The unrelieved jolting on a hard seat ensured that upon arrival I would be too travel-drained for this famous town, 'Gateway to the Gods', to impress itself favourably upon me. Indeed from where I stood, scanning this way and that, I couldn't see what was so special. A few yards away a bus raucously revved its engine, while the conductor leant from the door, chanting its destination, "Rishi – Rishi – Rishikesh!" A dog raised its leg against one of the dusty rear wheels.

I turned my back as a spurt of acrid exhaust fumes threatened to engulf me: Rishikesh could wait till morning. The late afternoon light had begun to soften the outline of the nearby Shivalik hills and I was in no state to hurry.

Previously known as Mayapuri, Haridwar has been a pilgrim centre for centuries and is one of the four places to celebrate *Khumba* every four years to commemorate a long ago victory of the ancient *Devtas* (Gods) over the warring *Asuras*, the Demons. The astrological date of these fairs, or *melas* – the largest being held in Allahabad when the visitors are numbered in millions – falls when the planets Venus and Jupiter coincide in January and early February. The venues are held to be the very spots where, during the great battle, a few precious droplets of the sacred *Khumb* (or nectar) spilled from a pot being carried by one of the deities during a chase across the heavens which lasted twelve days and nights. For centuries since, numberless pilgrims have made long journeys to these hallowed places; drawn by an irresistible, mysterious urge, they come and go in their thousands to Haridwar. Each morning and evening the river's edge is charged with their murmured chanting; fires are lit and the smell of incense hangs sweetly in the air. It is time to forget the petty affairs of the world.

I set off down the long, nondescript road, which I judged would bring me to the heart of the town. The distance as usual was further than it looked and I walked two kilometres before the road, narrowing, led me into a busy network of shopping lanes. Shortly, the lane's angle eased and about the same time I saw a promising – looking building with the river gliding swiftly beyond it. Hotel Raj, as I quickly discovered, was an attractive four-storey structure painted like the sign I had seen earlier, in bands of eggshell blue and white. My room was spotless and airy, its large window commanding a fine view across the Ganges to the wooded foothills of the Himalayas. Before the river and about thirty yards from my room a row of time-lichened domes formed a kind of balustrade over which my eye could linger upon the eternally flowing waters of India's holiest river. Greenish in colour, it rushed full and deep between the stone *ghats* and I sensed something of why it inspires so much reverence in the minds of millions of Hindus. For them it is truly the 'Mother' of so much which shapes their lives. Rather than raising such questions as, "Is it safe to drink?" I found myself thinking more of its healing properties, a quality widely attributed to its status as the manifestation of the Supreme Spirit. The scientific claim that its power to cure derives more from the many minerals which it gathers on its near- 2000 mile journey from Gangotri in the Himalayas to the Bay of Bengal on the eastern coast of this vast sub-continent. Sceptics, perhaps more familiar with the murkier rivers of their own countries, dismiss any such claims as no more than wishful bio-nonsense and reject the first explanation out of hand.

For myself, I prefer to keep an open mind on the matter. But perhaps times are changing and the long-held barriers of indifference to things 'Oriental' are crumbling. There are many signs of this being the case. My gaze shifted higher, to the distant temples which glowed like energised thimbles atop the Shivalik hills. Things happen in India which often sound fantastic to Western ears: subtle states of consciousness and super-physical experience – can they all be explained away as deluded thinking or hypnosis of gullible minds?

Possessions, all 'essential' to my survival in India, lay strewn across my bed. I would live 'simply' while travelling, I had casually assured myself while preparing to leave England. Each item – map, sleeping bag, folding *Opinal* knife from France, and the homeopathic diarrhea cures – each one was a supposed manifestation of this worthy philosophy, an extension of my notion of who I thought I was. But *simplicity* of living depends more on an attitude of mind than the number of *things* one happens to have. Throwing them all away doesn't necessarily make for a simple life. Simplicity – did I really know anything at all about living simply, with no desires or expectations or what tomorrow would bring?

After eating only three or four bananas all day, my stomach called for a more substantial fare, and I devoured a *thali* of rice and vegetables in a nearby sidestreet restaurant. Afterwards a narrow lane took me to an open space by the river, Harki-Pauri, the largest of Haridwar's bathing *ghats*. An object of special reverence is the *charan* or footprint of Vishnu, one of the noble Trinity of Hindu Gods, which can be seen imprinted on a stone let into the upper wall of the *ghat*. A month or two before, this area would have been choked to capacity with devout worshippers, food and souvenir stalls, as well as the growing numbers of curiosity-seekers. Tonight, I was glad to find that Harki-Pauri's spacious promenade held only a few hundred people. My main impression was of pastel-tinted shrines, shadows, the flickering of fires and the pale glow of oil lamps. The warm air vibrated softly with unobtrusive chanting, while from one corner came the playing of a stringed instrument.

A mood of great calm filled me, as I remained motionless for a while on the edge of the *ghat*. Soon I found myself drawn towards a gathering beneath a huge, spreading tree whose largest limb was supported by an equally sturdy wooden prop. The tree was floodlit and beneath it, an attentive throng was listening to an address by a Hindu monk. What he was saying I could not tell, though the rapt expressions he produced on the faces of many of his listeners reminded me of the gathering I had witnessed in Almora recently. In both situations the people were receiving no dry body of scriptural dogma. Instead, there was an almost tangible quality of the spiritual life being lived. When I moved away I felt a little purer, more whole.

I wandered this way and that – 'spiritual sightseeing' – until I found myself approaching a low stone bridge beneath which flowed the Ganges, aflame with reflected light. I crossed over into the semi-darkness beyond and sat down on the deserted steps close to the water's edge. Few people were about and for a while I sat in silence, staring into the now dark water and allowing the day's events to filter slowly through my mind. Not many minutes passed, however, before I became aware that a hundred yards upstream of the bridge, some kind of celebration was taking place.

There were lanterns, shadowy figures, voices and a general bubble of low-key excitement. I thought a *puja*, or religious ceremony, might be taking place and sat upright to await developments. The answer to my question soon came when suddenly, the invisible river beneath the bridge began to sparkle and dance with the hundreds of flickering flames of a 'light' ceremony. As the tiny lamps, each launched on its way by a separate hand, neared me, bobbing and dipping in the swift current, I saw how each fragile leafboat carried its own oil light resting on a cushion of bright flowers.

So populous is India with deities that there is scarcely a spare moment when one or another is not being celebrated. Each village has its own shrine or temple where, in the coolness of dawn the god is awakened with a lamp ritual, recitation of scriptures and music. The god's image is washed and anointed with *ghì*, or clarified butter, before being blessed with dabs of coloured powders and hung with garlands. Throughout the year numerous festivals are celebrated. In autumn the victory of Rama over Ravana is signalled everywhere. Díwàlí, the festival of lights, is acclaimed in October to November to mark the approach of the winter months.

I was struck by how the lampboats, some surging on far ahead of the rest, resembled our own brief lives on this planet. There were those which shone all-too brightly in the beginning, only to be snuffed-out at the first obstacle, their light doused by spray with life's journey scarcely begun; others disappeared, seemed lost, extinguished, only to ride up from a trough in the current's swell with luminosity undimmed. A few more confident than the rest rode the river with majestic nonchalance, shining with a steady, unflickering intensity until, at last, I lost them downstream in the reflected lights of the town. The whole event seemed to hum with a conscious reality which I found deeply moving. So absorbed had I been in the ceremony that it was a while before I felt the chill of a fresh breeze against my bare arms. It was time to return to my hotel. Midnight was closing in and I wanted to be away at a reasonable hour in the morning.

I felt re-energised by the riverside stroll, and instead of going straight to sleep when I reached my room, picked up a book, 'Illustrations on Raja Yoga', by Yogi B.K. Jagdish Chander which I had found in a bookshop during my short stay in New Delhi. Over the past few years I had practised meditation and yoga on a daily basis. Another viewpoint, especially from an Indian adept, could only serve as encouragement to keep going.

In England, it was some time before I overcame the popular misconception about yoga being *merely* a system of excruciating physical exercises. I didn't know that *Hatha* yoga was but one of several kinds of yoga: karma yoga, mantra yoga and *Jnána* Yoga, a form which follows the path to union with God through acquiring knowledge of the laws which govern the world. It understands that the main stumbling block to man's self-realisation is *avidya*, ignorance, which is overcome by developing its opposite, *vidya* or knowledge. The life of the follower on this path is divided into four stages or ashramas: the first ashrama is known as the *Brahmacharyashram* or student stage, the early part of life concerned with learning and education. *Grahasthashram* or householder, is that period of his life when the man works, is married and lives amongst other people. The third stage,

Vanaprasthashram (or forest ascetic) marks the time when he starts to withdraw from everyday worldly life and begins to act as guide and teacher to young people. In the fourth and final stage he becomes a *sanyasi* or hermit, owning nothing and relying on charity for his existence.

Bhakti Yoga or the Path of Devotion, doesn't follow the strict ways of *Jnána* Yoga but finds its way to God (*Moksha*) through worship and prayer. Women tend to follow this path more than any other. The Raja or 'Royal' yogi on the other hand, lays different emphasis on the term 'posture' or '*asana*', the bodily techniques applied in Hatha Yoga and its variations. The different yogas are all inter-dependent – no one method is 'the best' – but the Raja Way is more of an 'inner Yoga', involving 'spiritual' postures. More important than trying to balance his body on his head, is the striving to keep his head, i.e. his mind, in balance. Using subtle techniques he withdraws for a time from the world and the distraction of the senses (as a tortoise withdraws into its shell).

Opening the book at random, I was surprised to see some familiar lines from William Wordsworth staring at me from the page:

> For oft when on my couch I lie
> in vacant and pensive mood
> They (experiences) flash upon that inward eye
> Which is the bliss of solitude.

Some homely observations concerning psychological principles applied to everyday living accompanies the quotation. Even more intriguing are the captioned illustrations of Wordsworth. The poet, flashily attired in a sharp 20th century Italianate suit and red tie is shown strolling daintily along the shore of Grasmere, while a bank of the famous daffodils leans towards him as though to greet an old friend. The second frame depicts him at home, cigarette in hand, "reflecting in tranquillity". A memory bubble containing the previous scene hovers above the wondering philosopher-poet who now wears slippers and reclines on a sofa with IKEA chromium arm-rests and apparently carved from a large block of purple soap.

Interesting though Haridwar had proved to be, a strong inner feeling told me I should move on the short distance upriver to Rishikesh or 'city of seers' as it has long been known. The bus was empty and appeared not to have been swept for weeks, the metal floor being covered with a rolling carpet of spent peanut shells. Fifteen miles of shaded *sal* forest vanished into the dustcloud astern before we abruptly arrived at this renowned sanctuary for the spirit.

My first impression was not a flattering one I have to admit. I thought the town would be simplicity personified; instead it seemed squalid, overcharged with pollution and chaotically noisy. The bus blew an emphatic cloud of exhaust gases in its face as I disembarked, which seemed to set the seal on my initial reaction.

35

Not a solitary temple in sight. And where were the distinguished holy men, the monks and sadhus, faces aglow with knowledge of higher things? Where also was the Ganges which I had last seen soon after leaving Haridwar? To complicate matters further, I had come to the decision to change more money before I headed off to find an ashram willing to admit me. It would be Saturday tomorrow and I had heard that the banks would be on holiday on Monday as well. At such times one's rucksack begins to feel as if it is filled with cement.. Eventually, I found a sidestreet hotel in which it seemed I could change money.

The manager glanced me over in the lobby. "Indian holiday on Monday, sir – there will be no problem. Besides, I will lend you this money. You stay here – very clean room no monkeys – and I lend you. I don't know you, but I trust you." He seemed to know where I was heading: "When you get to ashram," he advised, "ask for Bobby – he will help with money. He has been there six months – everyone knows Bobby."

I picked up my rucksack, thanked him for his help and made a rapid exit; I would manage with what I had.

Near to the intersection I met a nonchalant German peeling a pink banana by one of the fruit stalls. About his neck and wrists he wore a variety of jewellery including a necklace of amethyst *japa* beads. No less affected by the clamour about him than the bony cows sifting through the nearby piles of rotting vegetation, he looked a good person to ask directions.

"'Swarg Ashram'?" he echoed, casually tossing the skin to one of the watchful cows. "Yeah, Swarg Ashram – it isn't really an ashram, it's a bridge, RamJhula. The ashrams, yeah, they mainly on the other side of der river, but not all of them. You can stay at those places…cheap rooms, food, meditation…peace." He swung me a charas-fuelled friendly glance. "I go in that direction now. I show you …half an hour only" – Exuding a self-contained air of whatever happened it didn't matter, he set off into the shimmering heat haze with me following. Underfoot was a slithery compost of trampled food and fresh cow dung.

I indicated one of the squawking, cockroach-like auto-rickshaws which swerved up and down the market in hungry shoals: "What about a taxi?"

A pained look crossed my new companion's face as he forged on into the rising dust. "Those things! It is much better to walk in India…Slowly, slowly."

"Easy for some to say that," I thought drily, in sweating pursuit, "especially when you aren't carrying anything."

After forty minutes of 'slowly, slowly' we came to a concrete bridge spanning a dried-up river bed."

"Swarg Ashram?" I queried shortly, staring blankly along the arid banksides of glaring boulders and scanty, sage-coloured scrub. My strolling guide shook his head. "Just a little way now."

We pressed ourselves to the parapet as an overladen truck careered over the bridge with a trail of pedal rickshaws, bicycles and more taxi-cabs in its wake. On the truck's rear was painted the polite request: HORN PLEASE. After another interval of trudging silence I asked the whereabouts of the post office. The response was a kind of self-absorbed monologue as though I had asked directions to the Egyptian Pyramids:

"Yeah, there is one somewhere, but I don't know where it is now. Seven years ago, when I first came to this place from Germany, I used to get mail. Today I don't send any and I don't get any. Not any more I don't." He paused to stare into a walled garden planted with shady trees. "Sometimes I go away for a little while – the south. But I live here. It's not so hard for us to live in India. Even when your visa runs out or you lose all your money."

"There's ways and means?" We were walking again.

"Yeah. In the smaller places – if you stay in the smaller towns, usually the police, they aren't caring very much. You give them a few rupees an' you can stay. If you want to work, there's always that chance of making a little money. Someone in the village might want to learn English, an' you can teach them: 'Yes'. 'No'. 'One'. 'Two'. 'Three'. 'Buckingham Palace'. Even when you don't know any, they think you do."

I laughed out loud for the first time since Ramnagar.

By now our shuffling gait had brought us to the limits of the commercial district; even the onrushing traffic grew lighter as, on the right side of the road, business premises gave way to open plots of land, crumbling stone walls enclosing solitary buildings and stands of tall trees. Here and there I spotted an occasional whitewashed dome rising above semi-tropical shrubbery. Things were looking up. Then a thrill shot through me when through a gap I saw the dashing waters of the *Ganges* for the first time at Rishikesh. Even from a quarter mile range, I could see that its surface was far more turbulent than at Haridwar.

My attention was immediately drawn to a building of curious appearance standing on the far bank. Painted in horizontal stripes of tangerine, custard and white, its rectilinear lines and several 'decks' suggested a surreal marriage of stranded cruise liner and an enormous block of Neapolitan ice-cream. The German caught the direction of my gaze and wafted his hand, Asian style:

"That place is cool…Ved Nikatan – a lot go there. Some ashrams, they don't allow foreigners. Others aren't very *shanti*: no smoking; no music." He nodded towards a wall of forested hillside rising beyond the ashram. "A woman from Germany has lived up there in a cave for thirteen years. It is hard to find her

37

place... a very small path. Sometime I call to see her. But so far I haven't. When the time is right."

He broke off as we reached a metal gate set in a high wall of whitewashed stone. Beyond was a garden and the entrance to a low dwelling about the size of a small bungalow. He raised a ringed hand to his light shoulder bag. "Now I have to go, a baba, he is expecting me."

With that, he looked at me steadily for a second or two before turning to the gate. He closed it after him and ambled away towards the building and did not turn round.

I trudged on.

For a few moments I missed the German's company – or any company. Then a young Indian boy, sweating as much as myself, and pushing a battered ice-cream cart, flashed me a warm smile which made me feel instantly better. Letting go – it is not an easy lesson to learn, whatever country you happen to live in.

Four, dusty hundred yards later, a prominent signboard by a flight of steep concrete steps announced: the *DIVINE LIFE SOCIETY*. This was Sivananda's ashram founded by the swami in 1936 and today the headquarters of a world-wide organisation, with branches in many countries. I had no intention of staying there, but my curiosity got the better of my tiredness and I mounted the tree-shaded stairway for a quick look around. At the top of the stairs was a spacious courtyard with stone seats, a variety of shrubbery and a few peaceful-looking Westerners strolling about in white cotton robes. In the background was a large temple, a library and a frontage of modest institutional-type buildings. No-one gave the latest newcomer a second glance as I sat down on a concrete bench for a breather and a pull from my water bottle.

The society's reputation far exceeds that of the 60s 'Beatles guru', the Maharishi Yogi, of whom more later. On a nearby stone obelisk placed so as to attract the attention of even the most casual visitor, were inscribed the founder's guiding principles: 'To speak the truth at all costs, to speak sweetly with love, to practice non-violence and continence'.

A booklet I picked up later told me the place functioned as a spiritual retreat for those 'who wish to refresh themselves physically, mentally, morally and spiritually'. One should also 'Serve, love, give, purify, meditate, realise, be good: do good'. A popular place with Hindus, too, it also boasts an Ayurvedic dispensary and an eye hospital with beds for 22 in-patients.

I didn't wait around long despite the relaxed atmosphere, and soon returned to the road and the world of feverish commotion. The temperature, a neon sign handily informed me, was a grilling 35 degrees. Almost immediately I caught a glimpse of a bridge superstructure and turned thankfully down a lane of whitewashed walls, three-table cafes and tiny souvenir shops, which sloped in its direction. Halfway down I paused at a juice stall where I bought a glass of freshly-squeezed orange.

Swarg Ashram bridge was a graceful, bow-shaped structure of slender uprights, wrist-thick wire ropes and a clutch of seated beggars at either end. A procession of chattering pilgrims passed me as I stepped on to it. For the most part, the men looked older than the women and were dressed in clean white *dhotis* and a white cloth wound turban-like about their heads. Though I noticed some striking faces amongst them, the women were much more eye-catching in their saris of shimmering peacock hues, while at their wrists silver bangles tinkled with every animated gesture. I had to sidestep a forlorn-looking man of no more than thirty propped against the bridge railings, one leg folded beneath him, the other, a gruesome sight, extended along the ground. He stared dully into my face as I approached.

It was not a situation one normally encounters in my own part of the world and my senses instinctively recoiled from what I was seeing: a large open gash ran the full length of the man's inner thigh; ending almost at the knee it was three inches across at its widest and looked like the sort of injury which might be inflicted by a shark or some sharp blade.

My feeling of shock was closely followed by one of indignation. The man was a clear hospital case, so why was he having to endure such a wretched plight? I heard afterwards that he was unable to raise the thousand rupees or so, which hospital treatment would cost. Feeling helpless I dropped a coin or two into his tin and carried on. His eyes however, were already engaging those of the next person and seemed barely to register my action. More spectral figures, one with leprous stumps for fingers, approached me when I stopped to look down at the water.

Like the timeless India of the travel brochures, it is easy to sentimentalize beggars, especially from the insulating distance of one's own country where poverty is a near-invisible phenomenon. In the deeply overcrowded towns and cities of the sub-continent, the poor are suddenly no longer the shadowy abstractions of a 'caring' intellect: you meet them face to face and they cannot be ignored.

Some Westerners decide they cannot cope with the poverty, disease and dirt and go home. Others conclude, from the viewpoint of ignorance, that India has nothing else to offer, and don't go at all. In Delhi and elsewhere I was importuned by mothers sometimes carrying very young children with wasted, mutilated limbs. Such injuries, I discovered, were often the work of background figures, the 'beggar master', who intended, thereby, to arouse the public to greater pity. How many are 'at work' on any given day in Delhi it is impossible to estimate. Thousands and thousands. They are as much a part of that city as the spacious tree-lined avenues, the luxury hotels and the stylish commercial buildings.

It was easy for me as a wandering visitor to ennoble beggars with their sometimes colourful costumes and quaintly old-fashioned English ("Excuse me, sir, could you spare a copper for an old man"?). The traditional view of Hindus, that beggary is a necessary demonstration of *karma*, the distorted limb being living proof of past misdeeds, will no longer do. "... the alarm has been sounded," says V.S.Naipaul in *A Wounded Civilisation:* "The millions are on the move. Both in the cities and in the villages there is an urgent new claim on the land; and any idea of India which does not take this claim into account is worthless. The poor are no longer the occasion for sentiment or holy almsgiving..."

I stared into the rushing water directly beneath the bridge. The resident over-fed fish, a shoal of some several hundred strong balanced as one body against the current's ceaseless tug, ever alert for tit-bits tossed from above. I would give to the beggars when it felt right, I decided, and for no other reason.

Three hundred yards downstream, where it swung to the left across shallows, the surface of the Ganges flashed as if a thousand jewels danced upon it. Mercifully, the town had now disappeared into a heat-haze of squat buildings, trees and drooping telegraph wires, the distant bray of a horn being the only sound to remind me of its presence. Rising high to the left of the bazaar and temple roof tops was a mountain ridge deeply indented with curtains of forest. I moved on, the sun beating down, the bridge carrying me, I sensed, towards a very different scene.

The glaring light hurt my eyes and by the time I reached the other side of the river much of my energy had drained away. However, the ashram could not be far off, and I continued walking, turning right past a couple of tiny music booths and a bespattered recess holding a Hindu deity. I stared into a souvenir shop without really registering its contents – a few Indian tourists were inspecting the gleaming ranks of ornamental effigies and the gaudy posters illustrating scenes from religious stories. Shops catering for the mind and spirit as well as the body caught my interest, as I proceeded according to the German's directions. Downstream was a succession of stone *ghats*, wide sloping platforms ending in steps leading down to the water. Being midday they were deserted. But at dawn, evening or any other auspicious time, they would be crowded with worshippers chanting, submerging and offering prayers as they performed their age-old devotional rites.

As I drew near to the dappled shade of the narrow bazaar, a seated, prophet-like figure with flowing white mane and beard swung about to face me. His features radiated charitable thoughts and wisdom – or such was my immediate impression.

The robed sage stared intently into my face and, raising a dramatic forefinger towards heaven, proclaimed solemnly, "ONE, SIR!"

His voice rang through my head as if he had uttered the Truth of all truths. I gave him an acknowledging nod, thinking he might be some market place Socrates, a dispenser of profound spiritual aphorisms. It was flattering to imagine that we two had something of significance to share, but I trudged on without bothering to find out. Short of a closer knowledge of such people, it is difficult to assess them with any accuracy. However, it would not be long before we met again.

After the uproar of Rishikesh town, the canopied tranquillity of the bazaar was balm to a fevered brow. An occasional scooter twisted through the few strollers, but otherwise there were no vehicles to jam the lane. The shopkeepers wore unruffled countenances and I also passed one or two figures clad in the ochre robes of monks.

As I emerged into the sunlight at the far end of the passageway, two Australians, almost invisible in the heat's shimmering glare, passed me heading the other way. I caught a snatch of an unsympathetic exchange concerning Rishikesh.

"There's not much going on in this place – just a few God-guys, temples an' stuff.."

"Reckon we should leave, tomorrow?" replied the second voice.

"Awrighty."

The Ganges, rushing on with impressive turbulence, now lay about fifty yards on my right across a boulder-strewn strip of sand. However, beyond a high wall and an open-fronted tea hut stood the unusual structure I had noticed from the other side of the river, the nearest wing of the building being protected by a wire fence atop a knee-high wall. A procession of wire screen cages containing several female deities presiding over strange, fearful-looking doll-men, led me to the simple metal gate. Beyond this a broad flight of steps rose to the ashram of Ved Nikatan.

It was a low two-storey edifice divided by twin-doors and wire mesh windows, into cells or individual rooms. The roof and the floor of the second level were supported by stone columns painted a dazzling tangerine. A line of washing above the balcony was the only sign of life. At the highest point of the ashram – a three-tiered pyramid supported on orange columns, hung a limp red flag. Another flew from a pole overlooking the main entrance. Founded earlier this century by Yogi Sri Vishwaguruji Maharaj, the policy of Ved Nikatam is to accept all visitors whatever religion they happen to follow. A hand-painted notice informed visitors about the ashram's function.

It was: *A Centre of Universal Education. Learn Good Teachings, Sacred Seed Formulae, Virtue-Based Meditation and Easy Raja Yoga.*

I could see no immediate sign of anyone pursuing these noblest of goals, only a young foreign man dressed like a sadhu, peacefully asleep on a stone sofa. Behind the slumbering Westerner, a further two notice boards attracted my attention. Neatly hand painted, they were filled with the conditions of entry and austere rules of conduct which promised to vigorously examine the hidden weaknesses of all who set foot within.

The *Hare Krishna* devotee's pronouncement that ashrams had "a more serious purpose than hotels", floated into my mind as I scanned the list with furrowed brow. It would be a waste of valuable paper to note all the conditions of admission (there were nearly 30); a token sample should be enough to convey the flavour of the kind of regime which awaited the would-be ashramite.

> *Residents Must Take a Decision to Adopt Virtue and Give Up Vice.*
> *Drinking and Playing Musical Instruments is Not Allowed.*
> *Tourists and Critics are Not Allowed Entry in the Ashram.*
> *Making Cutting Jokes, Undesirable Gestures or Reading Dirty Novels is Not Permitted.*
> *Inmates are Not Allowed Visitors in Their Rooms After 6p.m.*

Having slogged all the way from town, I was suddenly filled with foreboding and paused for some time before the unattended entrance. Should I or shouldn't I? Surely only a product of the English education system or someone completely off their head, would willingly enter such a world.

I decided to try it.

The foothills of the Himalayas at Rishikesh, with the Ganges just seen centre left

* <u>CONDITIONS OF ENTRY</u> *

1. Use of onions, garlic, meat, fish, eggs, hashish, bhang and cigarettes is strictly prohibited,

    *

2. Playing cards, chess, chowser, political discussions, cutting jokes, undesireable gestures, and reading dirty novels, is strictly prohibited.

    *

3. People staying in this ashram are not allowed to keep any person staying with them without permission.

    *

4. Devotees should wear yellow or plain clothes, ladies should not wear costly ornaments or foreign dress.

    *

5. Tourists and critics are not allowed entry into the ashram.

    *

6. No outsider is allowed to enter the room of any pilgrim without permission from office.

    *

7. People staying here should look after the safety of their belongings themselves. Management will not under any circumstances be responsible for any loss or damage.

    *

8. Do not sit without purpose.

    *

9. Upkeep of sanitation is the sole responsibility of people staying in the ashram.

    *

10. Do not feed the monkeys.

"I paused for some time before the unattended entrance"

Chapter 3

The Floating Stone

I passed through the gate to an airy, deserted reception area with two open doorways immediately opposite each other. In the centre of the hall stood a metal tank half-full of brackish water in which wallowed a semi-buoyant object which reminded me of a very old bathroom sponge. A notice invited the visitor to *View Floating Stone.* The left-hand door opened into a dusty vestibule containing various Hindu votive objects and a collection of faded, yellowing photographs tracing the ashram's history. Many of these were of the present guru *Sri Vishwaguruji* and showed him officiating at various religious celebrations or in more informal situations, posing in a variety of Hatha yoga *asanas*. A pamphlet told me that *Guruji* had been initiated into the holy order of *sanyasins* by his guru, *Sri Sri Hiranananda Saraswatiji Maraj* who had lived to be 125 years.

Under his guru's guidance, *Guruji*, the pamphlet said: "practised rigorous austerities, lived in jungles and caves, survived on roots and fruits, travelled extensively on foot to all the sacred shrines of the Himalayas and became an adept in the science of Hatha yoga, Kundalini yoga, Gayatra mantra, sadhana and Raja yoga meditation".

The *Guruji* has about 40,000 monks under his training as well as many hundred thousand lay disciples and devotees. Later I discovered that almost his only English was the sentence, "I like you".

From within a door marked 'office' came the whirr of a fan and the sound of steady, rhythmical breathing. Poking my head inside I saw an aged, hollow-cheeked monk slumbering soundly in a reclining position amongst a disorder of loose cushions. His mouth was partly agape and I saw he had very few teeth. Before him on a low table were scattered several scriptural texts as well as a couple of well-thumbed ledgers – the spiritual and the worldly juxtaposed.

I coughed, but got no response. After a pause I tried again with the same result. The ancient twitched, shifted his position, and slept on.

What was I to do now? There was no bell for attention. The ashram was supposed to be a centre of awareness, yet the only two beings I had spotted so far were both sound asleep! I turned away from the office, gave the 'floating stone' a less charitable inspection than previously, and wandered unaccosted to the wide inner gate which opened on to a silent quadrangle of bleached grass.

The ashram comprised a high-walled compound divided into two courtyards of unequal size by a colonnaded flat-roofed block of rooms with brightly painted doors. The block was split by a bow-fronted, two-storey temple whose stepped entrance faced directly towards the inner gate. On the temple's left hand side, a whitewashed alley gave a glimpse of the smaller courtyard beyond and yet more rooms with barred, gaol-like exteriors on two floors, The place was obviously intended to cope with a fair number of inmates.

But where were they? The hushed atmosphere may have had more to do with the soporific heat than anything else, but I found myself wondering if anyone was behind those firmly closed doors. I sat down on the top step and reached for my wineskin water bottle, given me by a French woman in Trieste years ago.

A generator's muffled *dub-a-dub* somewhere behind the temple was the only sound; a splay-legged calf tottering before a dry tap was the sole sign of life. Yet after the mayhem of town or even the placid neighbourliness of the nearby bazaar, it was unnerving to be so abruptly confronted by such silence. The closed doors returned my gaze with studied uninterest; overhead, too, the sky was an unheeding hemisphere of drained colour, a glazed arch of infinite distance and vitrifying heat which, rebounding from the whitewashed walls of the compound, took pity on no-one and sucked my throat dry. There was no respite – unless one was fortunate enough to have a room. Ah, a room with a fan and a bed to lie on! I scrutinised a complicated crack-system in the ceiling for a while then, with renewed determination, strode back to the office.

In my short absence a miraculous event had occurred: the monk was no longer lolling in the land of Morpheus, but semi-awake, and peering at one of his scriptures through a pair of heavy, black-framed spectacles. His hand shook occasionally and if he noticed me, he gave no indication of the fact. Only when I stepped inside and sat down on the worn mat before the small table did he look up:

"You want to stay in ashram?" he addressed me in a short, high-pitched voice.

I nodded, holding up three fingers: "*Tin* – I stay three nights." I raised my voice in case he was hard of hearing.

The monk struggled down from his cushion clutching a maroon visitors' book. "You sign…passport number.. Country."

A staccato babble of Hindi and English bruised my ears as I filled in the relevant information. Against 'Destination' I wrote, 'Gangotri?' When the old man mentioned "rupees" for the second time, I fished in my money belt and handed him the last of my paper money, a hundred-rupee note.

He took it in a long, knotted hand, inspected a half-inch tear in its central fold and called out some further directive through a doorway to his left. Time dragged with exasperating slowness as he rummaged here and there until, finally, he began to write out a receipt. I folded my hands and tried to be patient. Was this the ashram guru? My eyes swung to the key board. Would I get a room, one with a shower? There was no way of knowing.

At last the task was completed and he handed me a slip of paper, adding enigmatically as he did so, the words, "Dharmananda Monday."

He called through the door with increased urgency.

This time the sleepy figure of a much younger man, wearing a white singlet and *dhoti* stumbled into view. A few short words passed between him and the old man before the former plucked a key from the wall, indicating that I should follow him.

Halfway across the quadrangle I remembered the hotelier's words:

"Bobby, he here?"

My guide gave a blank grin, mumbled a phrase or two in the vernacular which I did not understand and lapsed into silence. India has the kind of climate which does not encourage persistence and I didn't bother to repeat myself. "Bobby" was obviously a figment of the hotel manager's fertile imagination.

The room's twin doors, painted a jungly blend of green and banana, swung inwards on un-oiled hinges across a pitted stone floor. My guide pressed his palms together in kindly fashion and disappeared without a sound.

It was a dusty, traveller's twelve by ten, a lonely dwelling if you couldn't handle your own company – or the quiet. The furnishings, too, are worth a mention: a waist-high cupboard was the only feature of the plain wall to my left; facing it from the opposite wall were three cobwebbed stone shelves painted bird's egg blue. Both the floor and the barred window ledge were coated in a film of dust. A wire-screen window promised to repel all but the tiniest insects and a pair of outside shutters offered further seclusion from the prying eyes of the outside world. Illumination was provided by a solitary unshaded light bulb dangling from the ceiling on the end of a twisted cord. My room was certainly frugality personified for there was also no bed.

I started to walk about, inspecting details here and there which had escaped my notice. Of recent occupancy the room bore little evidence, though someone had once used the topmost of two stacked bricks in the dimmest corner as an ashtray – the crumpled remains of a *beedi* and its scarlet binding thread told of another who had stopped by, and moved on.

I flung open the cupboard to ensure it was not home to some unpleasant, sub-tropical crawlie. I was in luck, for apart from an old incense cone, the cupboard, as they say, was bare.

The ceiling fan had two-foot blades and a three-speed control knob which looked promising. I tried it at HALF and finally FULL only for the whole unit to waver dangerously in its bakelite housing just a foot above my head. I turned it off. At the rear of the room a second pair of doors opened into a narrow brick yard containing a wire washing line, shower and Asian-style toilet with its ubiquitous can of rusty water. Not a place to linger. The generator's steady *dub-dub, dub-dub,* was closer now.

Ashram rooms occasionally display interesting Siva-inspired wall art, some of it dating back to the 1960s flower revolution. I had obviously been given the scribblers' gallery for my pains, for here and there was a scrawled line or two of borrowed wisdom: "Every journey begins with the first step"; "Wherever I am it is now". And so on. On the wall opposite the window was a wavery *OM* in pencil. I can sleep pretty well anywhere if the occasion demands, but I'm sure that not everyone would share my apparent enthusiasm for such homely circumstances, even for less than one pound a night. However, as I washed my cares away beneath the tepid shower, I was grateful for small mercies.

Later, finding a twig broom in the backyard, I used it to good effect on the bare floor and any other surfaces within reach. I spread my lightweight groundsheet up to the wall and unrolled my sleeping bag along it. To provide some insulation from the hard floor I folded the several items of cold weather clothing and laid them beneath my zip-up bag. I confess to being a little surprised to have no bed of any kind, but its absence seemed in keeping with the notice board commandments on the outer wall, and I gave the matter minimal attention. If that was the way of ashram life, I would just have to adapt.

Not willing to startle myself again by turning on the fan, I picked an orange from my newly-designated food shelf, and sat down on the outside steps.

There was now no breeze at all. The still air hung like held breath in the quadrangle; the silence, the inactivity and the staring rooms oppressed me. Perhaps I was just tired? Above the roof-top the flag hung limp against its pole. I felt consumed by the desire for an ice-cream. As soon as I finished my last piece of orange I felt thirsty again, and reached for my water container. Across in the reception area a few figures, enigmatic, insubstantial beings flickered silently to and fro. They looked unreal, as though they belonged to a long-forgotten movie with no name. And for a few hours, not without good reason, so did I. After half an hour, the heat forced me back to the shade of my room, where I lay down full-length beneath the fan with the control turned to HALF. I checked my watch. It was only 2 p.m.

Once, hearing a desultory chord or two of 70s guitar music, I sprang to my window. Too late! The strumming, which came from one of the end rooms, finished as abruptly as it began, as if the player lacked the energy to continue.

The only event to break the silence of the next two hours was the harsh sound of a bolt squeaking open on somebody's door. This time when I peered out, I was rewarded by the sight of a fair-skinned forearm, and a banana skin flying into the empty quadrangle. A moment later the bolt rammed shut. The silence returned as before, pressing down upon the place like the wings of some alien bird which the world knows only in sleep or in the painted visions of some half-mad artist. A bead of sweat trickled from under my armpit as I dozed off again. When I next glanced outside the banana skin had vanished.

I should hardly have been surprised by the fact that no-one was about- after all, ashrams, retreats and other such refuges from everyday bustle are intended to provide space for those seeking inner quiet. My own mind, though, was still much in its outgoing Western mode and needed to drop a gear or two, cool out, before I could stop feeling I was somebody going somewhere else. When you begin to feel that whatever is happening right now is going to turn out all right then you're on top. Staying centred is what it's all about.

I finally stirred around five o'clock. Nothing seemed to have happened, but when I turned my gaze towards the window, I noticed the light had grown softer as the afternoon crawled towards the distant Bethlehem of early evening; some of the sharper outlines of my room's interior had begun to melt, dissolve into mere shapes within shapes and places of ambiguous shadow.

How slowly – yet how suddenly time had passed! I now had a little more money, having discovered, whilst idly flicking through my passport on a foreign visa count, a folded R50 note which I had completely forgotten about. I now had no urgent need to visit a bank or, for that matter, to find the elusive "Bobby".

After a few minutes, I reeled to the shower where I doused my head briefly to freshen myself up before heading back to the bazaar.

Mercifully, the fiery heat of early afternoon was now more bearable and I paused on the ashram steps to survey the peaceful view which greeted me. An avenue ten feet broad, marked by parallel lines of whitewashed stones had been cleared through the fifty yard strip of boulders which separated the ashram from the Ganges, now a flood race of molten silver flecked with carmine. At the water's glittering edge, several motionless figures stood or sat in attitudes of silent contemplation, their still forms scarcely separable from the boulders about them. Already the tip of a temple over the water flared with evening light, as another timeless day drew to its lingering close. Some yogis have left Rishikesh for quieter regions in recent years, but for those who have the time and patience to seek it out, not all of its virtues have been swept aside. I was soon threading my way through the shaded recesses of the bazaar's small shops and stalls, smoky cafés and beckoning temples.

48

Sharing the lane with me was a variety of leisurely humanity: old women with leathery faces and shrunken arms, beggars, both male and female squatting by their collecting tins; an Indian family or two on holiday, gazing at the trinket displays; gaunt ascetics with cheeks like saucers and wearing little more than *dreads* and an absorbed air of other-worldliness: all these and more thronged the narrow way. The colour of the scene was further enhanced by those pilgrims bearing garlands of golden marigolds about the entrance to a temple; some muttered sacred syllables as they passed back and forth through the gate, while others fingered their rosaries.

It was like nothing I had experienced in Europe and I felt quietly intoxicated by the sense of expectation which seemed to fill the air. I proceeded on my ambling course until I found myself before a bookshop specialising in spiritual and abstruse subjects. The sole occupant, the proprietor, a middle-aged man, sat musing quietly to himself on a rug-covered dais. At the far end of the shop was a glass kiosk inside which a youth was raptly watching television. The bookseller affected not to notice me as I entered the place and I was left alone to finger the various volumes which I pulled from the dusty shelves.

I hadn't come to India merely to wander the intellectual maze of arcane religious tomes, nevertheless my attention was soon drawn to a name revered the length and breadth of India, *Sri Ramakrishna*.

Born in Bengal in 1836, the sage never lectured or wrote anything down, but communicated his profound teachings in the form of conversations, a large number of which were recorded for posterity by his many followers. Preaching no particular creed, philosophy or dogma, he eschewed the formality of sermons, relying instead upon illustrations from nature and the simplicity of his own daily life to transform others, and to give them insight into the nature of the world and their own selves. Different temperaments and varying levels of intellectual capacity amongst his listeners were no barrier to his ability to touch their deeper natures; he had the uncommon ability to meet them all on equal terms. One minute he might be singing devotional songs or chanting *japas* (the repetition of sacred sounds or mantras), the next might find him explaining a tricky ethical problem, such as the perfection of God in a suffering world.

I opened the book at random to *Who is a True Teacher?* I read *Saying* 185:

"When corn is measured out to a purchaser from the granary of a big merchant, the man engaged in measuring out goes on unceasingly with his work, having a constant supply of grain. A petty dealer's store, on the other hand, is soon exhausted. Similarly, it is God Himself Who unfailingly inspires thoughts and sentiments in His devotees, and that is why they are never lacking in what is new and wise. But the book-learned, like petty grocers, soon find themselves short of thoughts and ideas."

Daunting stuff. It is like a child mistaking a kitchen tap for the source of water, when in reality it comes from the sky. I returned the book to its place and moved on.

Across the way from the bookstore was a herbal dispensary selling all manner of cures, some promising more unusual results than others.

I scrutinised the smudgy label of an oblong cardboard packet. *Formula No. 12* it read, *A General Tonic for Mental Imbalance.* The cure of what sounded like a very Western complaint was to be taken *when people begin to tear clothes, hurl stones at each other, talk at random, keep mum or mumble to themselves.* I recognised one or two of the symptoms in myself and rapidly picked up another item. For only R3.50 I would be *relieved of blood disorders, leprosy and itch.* In the same tone of absolute confidence, a third preparation, a whitish powder, promised to *stiffen loose breasts* after *forty days'* application. I found myself wondering if it was a kind of slow-drying plaster of Paris.

"Hair dye?" offered the owner, glancing at my not so young beard. I passed on empty handed.

Dusk had thickened appreciably by the time I returned to the vicinity of Ved Nikatan. The flanks of the hills, at midday, humps of khaki and shades of grey, were now curtains of darkest green, their tops lit faintly by the sun's slow afterglow. Across the Ganges, whose hue had now darkened to a deep blue-green, a small flame shone brightly, flickering now and then against the rising breeze. I stood spellbound for a full minute before this, the profoundest of scenes, until a nearby loudspeaker began to croak the scratchy, overlapping cadences of the Hare Krishna mantra.

A few more inmates, most of them foreigners, were now astir around the ashram, but I had still not spoken to anyone. A naked light was burning between the shutters in the room next to mine as I let myself in quietly, anticipating another solitary night with only dreams for company.

Scarcely had I dropped on to my sleeping bag, however, when my door began to shake beneath a tattoo of thunderous knocks. I of course knew no one in Rishikesh and wondered who it might be. Hoping to God it wasn't a summons to some monastic vetting procedure or an all-night meditation, I reluctantly slid back the bolt.

In the pool of light which fell from my doorway, stood a total stranger. An Indian of medium height, he was in his sixties and wore a homespun woollen jacket over a white *lungi* (calf length cotton skirt) and *kirta*, a shirt of the same material. Beneath his turban was a grey-stubble chin and a visage which made me think of desert assassins.

Giving me no time to utter a word, his eyes darted fiercely to my face: "You eat with me – next door! Meal nearly ready!"

Without waiting for a reply, he turned and disappeared abruptly into his room, though leaving the door ajar.

I didn't take kindly to being addressed as if I had no choice in the matter, but after a few moments useless prevarication, politeness (and curiosity) won out and I duly closed my own room and entered his. I would surely return within the hour, I told myself.

There was a pronounced nip in the evening air and at first I was agreeably surprised to find that my bluff neighbour's orderly room was a good deal warmer than mine. It was also far more comfortable, the floor being entirely covered with rugs, albeit frayed ones. A number of cushions were strewn about, while a thick mattress was rolled up against one wall. Before the bedroll was a low table piled with books and writing materials. In the far corner was the reason for my invitation – a metal pot simmering quietly on top of a smoky paraffin stove.

"Sit!" my genial host commanded, closing the door behind me and gesturing matter-of-factly towards the vintage saucepan, "eat twenty minutes."

I did as I was told, lowering myself onto a couple of cushions to await developments.

A few seconds later, a youth of about 16 appeared, wearing a stained *lungi*, a chest-hugging singlet and a curious grin which I never saw leave his face during all my weeks in the ashram. His smile widened as he squatted amiably by the bubbling pot. I wondered if he might be a bit simple but it was difficult to tell, since he never spoke more than a couple of words in my presence.

The 'major' as I quickly dubbed him, made no effort at conversation. This puzzled me a little, but feeling tired, I was happy enough to sit in silence for twenty minutes as my chin sagged gently and inexorably towards my knees.

Suddenly, just as I was about to keel over asleep, *the* question split the air: "YOU ARE FROM!?"

"Er, England," I mumbled thickly after an interminable pause for thought.

He snorted to clear his nose. "I am Kashmiri business man."

"Srinagar?"

"Yes, Srinagar. I stay here four months every year." His jawline relaxed for the first time and my impression of him as a fractious hill-tribesman softened a micro-fraction.

"A long way, Kashmir?" I said, as the boy handed us a chai apiece in metal cups.

The Kashmiri nodded, sipped noisily at the scalding liquid and lapsed again into silence. Only when we began to eat nearly an hour later, did he open his mouth again. It was the usual fare of rice and *dal*, though on this occasion accompanied by a deep bowl of bullet-hard uncooked beans or pulses which my neighbour flung into his mouth by the handful. Judging from the sounds which accompanied the action, he must have had teeth like a quarry crusher. I took three, perhaps four, cracked one gingerly between my molars – and swallowed the rest whole. The last thing I needed was a visit to a small town dentist with a broken tooth. I tried to look as if I was enjoying myself, but the Kashmiri had the eyes of an eagle.

"No!" he exclaimed, "not like that. You are not supposed to eat them *one* at time! You must eat. I am sixty years old, twice your age! It is good for you to eat – like this!"

With that, he tossed a further palmful into his mouth with unerring accuracy, giving me another demonstration of what sounded like broken glass eating. My fillings felt too fragile to withstand any more punishment and I waggled one hand in refusal; I hadn't skimped on the rest of the food and indicated the fact with a 'stomach full' mime, which seemed to satisfy him.

Changing the subject, I indicated the contents of his table.

"You are on pilgrimage, *yatra*?"

The Kashmiri nodded shortly. "I meditate six hours every morning."

"You sit six hours!?"

"No, not like that," he continued, in his gruff but understandable English. "At 4.30 I get up. I take shower. Prayers… I go to Ganga. Sit. Walk…read. Like that. I open my door at 10 o'clock."

He finished his chai and fixed me with a severe look:

"How long you meditate?"

"One hour," I claimed, which was fifty minutes longer than some people I knew.

The Kashmiri looked unimpressed. "What time you get up?"

"Five-thirty," I replied virtuously. "Usually."

He wagged a cautionary finger. "You must do much more; it is not enough."

I could have responded irritably, but his tone was less harsh than before the meal; a more companiable light now glinted in his narrowed eyes: obviously a better man for knowing *after* eating.

However, mainly due to my own increasing desire for bed, this potentially interesting subject rapidly tailed off. My own shortcomings in the matter could wait till another time, especially since the Kashmiri seemed to see it only from his point of view. As the boy began to clear up I sensed this was my opportunity to make an exit while the going was good.

I stood up to go.

As I did so, he took me by surprise, swinging suddenly about and seizing me by the wrist. I thought for a second he wished to restrain me from leaving but his intention was merely to inspect my wristwatch. Satisfied, he dropped my arm and pressed both hands to the side of his temple, miming sleep.

I nodded only too gratefully. Midnight was only an hour away and I had eaten my fill of both food and my host's testy injunctions for my self-improvement. Rest, not more mental browbeating was the answer.

Within two minutes of leaving his room I was fast asleep in mine. The night's interruptions were not yet over, however.

At some time in the early hours – I am not sure when – I awoke suddenly in the draughty darkness. My forehead felt clammy whilst the blackness was so intense I could scarcely see my hand before my face. A recollection, a memory of a dream started to twitch in my brain.

In my dream I was alone in a forest. A lion popped its head round a tree and stared at me with large, round eyes. It began to blabber in an archaic form of my own language.

A dim intuition told me I was being warned of some personal danger by the dream, but as to its nature I had no notion at all. All at once the lion seemed to lose patience and sprang out into the open. I shot up the nearest tree and awoke instantly, listening for suspicious sounds. A gusty wind skirled a pattern of dry leaves against my door, but it was not the night breeze which had disturbed my rest.

Suddenly I heard it.

From behind my pillow issued a furtive scratching noise. In a flash I sprang from my sleeping bag and snapped on the light. As I did so a six-inch millipede emerged from its hideout behind my folded pillow. Sensing the light, it pressed itself motionless to the foot of the wall. My sluggish state cleared in a moment as I reached for the nearest implement to hand, a flip-flop, with which to settle the situation. I was taking no chances after a young Danish traveller told me how he had been stung by such a creature, as he pulled on his boots in a Dharamsala hotel. His lower leg had swollen to almost twice its normal size before the intervention of a Tibetan doctor put matters right.

At one time my instinct would have been to end the drama by abrupt intervention with a heavy object, but in recent years my attitude had begun to change. The millipede hadn't set out that night with the intention of doing me harm. If it did sting me, it would surely be no more than an expression of its own wish to survive a stressful situation. The room was not only _my_ territory. Using my flip-flop, I flicked it across the floor, where it landed at the very lip of the drainage hole from which it had probably emerged. After a pause, it vanished from view. In case it had second thoughts I pushed the brick ashtray hard up against the opening. After that little escapade it was some time before I fell asleep for the second time; only after I had checked around carefully for others of its kind, did I lie down with a peaceful mind.

At last a grey light at my window signalling the end of my first night in Ved Nikatan began to chase the shadows from the corners of my room. As I turned first on the one hip then the other the generator spluttered into life in the rear courtyard. After a few hesitant strokes it settled into a steady *chink-a-chink* rhythm which was almost mesmeric. Nearby I heard a door close softly, whilst over the ashram walls a loudspeaker system cleared its throat before breaking into the Hare Krishna chant. Caught by the breeze, sometimes the words drifted across the quadrangle, sometimes fading far away, before loudly returning in a gritty burst of very unmystical static. I lay on my back staring at the ceiling for some time, reluctant to admit that my interests would be better served by a dawn visit to the *Ganga*, than by a vain struggle for an extra hour's sleep. A few further minutes passivity followed before I roused myself and pulled on my clothes.

The word 'ashram' derives from two Sanskrit sources: *shram* meaning 'hard work' and *ashraya* meaning 'retreat' or 'place of refuge from the world'. Their purpose is to provide a place of security, a community, for those who wish to pursue spiritual goals in peace. Some are closed to foreigners, but others are not very strict about a Westerner's spiritual credentials. Just what kind of household Ved Nikatan was in practice, I leave for the reader to judge.

The *shram* aspect of ashram life was conspicuously absent as I made my way back to the exit which was already open. For those unfamiliar with the terms, such notions may conjure up a Hellish scene of ceaseless toil in some windowless dungeon of forced labour – all on a measure of rice and a prayer, not to mention the sub-continent's enervating heat. To pay for one's stay and then agree to a few hours work each day, may seem an odd idea to the uninitiated.

As I padded down the damp avenue of painted stones towards the river, the funereal callings of crows mingling with the Krishna mantra produced a curious homage to the new day. One or two stars lingered on over Rishikesh town, whilst behind me, the backcloth of hills was still a drapery of folded darkness.

Other swathed figures besides myself were also paying their quiet respects at the riverside. Not far from me sat a fair-haired foreigner, a woman of around thirty. Dressed Indian style in a white sari of unembroidered gauze, her face was scarcely visible as she dwelled in her own thoughts amongst the pudding-like stones. We would pass one another in future days, but so enclosed in her own world did she seem, that she noticed neither me nor anyone else. Further on sat my neighbour, the Kashmiri. How long he had been there I could but guess. The cold water stung my bare legs, quickly dousing the remaining dregs of sleep. The swift current exerted a powerful tug, but I waded in determinedly until the water reached my rolled-up trousers at knee height, prudence telling me that to go deeper was to risk losing my footing on the stony river bed. It was too early yet to think of looking for breakfast in the bazaar, so after a token wash in the near-glacial Ganges I retreated temporarily to my room, where I soon turned my attention to a book I have found to be of unflagging inspiration on matters mystical: ' The Way of the White Clouds', by Lama Anagarika Govinda.

Govinda was one of the last Westerners to journey through Tibet before its invasion by China in 1950. With his ability to communicate a poet's response to a unique landscape and a true pilgrim's sensitivity to the ancient Buddhist traditions, it would be a dull mind which failed to be uplifted by his tale of travel amongst places of deep spiritual resonance and mysterious powers, which we in our 'advanced' Western society call 'miraculous'. The Tibetans think of them as normal. Clearly he was a participant as well as a wanderer.

Of the individual's need for silence he is in no doubt:

"… the absence of the spoken word, the silent communion with things and people, which was forced upon me due to the lack of a common language, can bring about a deeper awareness and a directness of experience which generally is drowned by the incessant chatter under which human beings hide their fear of meeting each other in the nakedness of their natural being."

He also distinguishes between a pilgrimage and an ordinary journey in that the former does not pursue a predetermined plan or fixed itinerary: "…it carries its meaning in itself, by relying on an inner urge which operates on two planes: on the physical as well as on the spiritual plane. It is a movement not only in the outer, but equally in the inner space…".

The Krishna tape crackled to a close just as I headed outside for breakfast.

Three foreigners, a Frenchman sitting with two cold-looking women, were the only other customers in the dingy restaurant halfway down the market lane. At the front, the owner was coaxing a reluctant paraffin stove into life. He mumbled a greeting as I sat down towards the rear, which looked out across the river. After an interminable delay my order of scorched sliced bread, dignified by the appellation, 'butter toast', and a chai arrived, delivered to my table by a young boy who could not have been more than twelve years old. I glanced over to the other table. One of the women had incipient, matted dreadlocks and wore a silver stud near the tip of her retroussé nose. It wasn't such a warm morning and her fingers were enlaced around her chai glass. I could see goose - pimples on her tanned forearms.

"Ved Nikatan?" I ventured.

The Frenchman's companions seemed not to understand, but he answered affirmatively in friendly enough fashion. Encouraged, I mentioned there didn't seem to be a lot happening at the ashram. Fingering his short moustache, the Frenchman replied,

"Sundays, yeah....nobody moves. The swami, he 'as been away. Tomorrow you will see him, Dharmananda." His words began to drift as though concentrating was too much: "The others, they speak 'ardly any Engleesh."

I told him I'd noticed, but I could see I was in taciturn company and didn't attempt to press the conversation further.

I lapsed into my own thoughts, sipped a second chai to wash down the rest of the toast, and left. I didn't feel up to a long walk, but the glow of light beginning to warm the riverside rooftops quickened my interest and I decided to explore upstream for a while. For the next hour or two the air would be free from haze, invigorating and hinting that the regions of the high Himalayas were now almost within reach.

It was wonderful to walk so lightly without provisions or luggage of any kind, except for an apple and my ubiquitous wine-skin. The only sound to disturb the tranquil scene was the occasional swish of my own clothing against passing bushes. However, I later heard that the way was not as peaceful as in days gone by, and a number of Westerners had been robbed whilst taking lone walks there. By day, though, the route seemed innocent enough and I continued to frequent it. My thoughts were to push on as far as the next crossing, Laxman Jhula, but soon after skirting a thorn hedge protecting a sadhu's dwelling, I spotted an opening on my left leading down a boulder embankment towards the river. It was too good an opportunity to miss.

Slabby volcanic rocks followed by smooth sandy hollows sparkling like diamond dust in the transparent air, quickly brought me to a perfect spot for a ten-minute dip: a sloping crescent of sand leading gently down to the river with a protective wall of black volcanic rock behind, honeycombed by numerous cavities. In this part of the world they are just as likely to harbour insects as interesting crystals and I resisted the temptation to put my hand in them. The sun was already warm on my neck, as I stripped off my light clothing, and dived in.

I returned to Ved Nikatan by the same route, stopping only to spend a few rupees on some fruit, which would last me the rest of the day. By the time I neared the bazaar, the harsh energy-draining light of approaching midday had already robbed it of much of its earlier charm. I had been out longer than I intended and the quadrangle, too, was awash with the same blinding glare which had greeted me on my arrival the day before. I sat down on the top step using Govinda's book as a cushion and took a long pull from the tepid contents of my water container. Over to my right, two or three Western visitors lolled beside their open doors, reading or simply lying motionless in the shade. The calf sat beneath a sheltering wall, its tail spasmodically twitching to ward off flies. No sharper contrast could be conceived than that between this tranquil courtyard and the exhausting, hectic scenes which I had so recently experienced. I smiled into the rectangular space without any conscious reason why I did so; but I didn't care. My idle ruminations were shortly interrupted by a light footfall behind me and an intake of breath.

"'ello. 'ave you just arrived?"

I half-turned to see a woman of Mediterranean looks staring down at me, as though eager to talk. Small, dynamic and on the right side of thirty, she was clad in a pair of candy-striped cottons and open-toed slip-ons. She met my querying look with an instant stream of effervescent bonhomie which should have rung warning bells, but didn't. Her name was Antonella.

"Tibet," she gushed, her eyes pinned to the tell-tale book protruding from under me, "it is almost like the moon!" She threw her hands in the air. "Almost you are on another planet when you are in that country," she bubbled, wide-eyed and earnest. She sighed theatrically as if to convince me that only Tibet had it- and she had been there!

For a while I was carried along by her flattering attention and air of conviction for all things Tibetan. After all, I too wanted to go there in the not too distant future, and a first-hand account was likely to be more topical than reading books on the country. Before long though, a more cryptic side of her nature began to surface: I almost needed a snorkel and mask to stay afloat on the Italian's deluge of travel anecdote, flimsy acquaintance with Eastern religions and a stream of off-beam generalisations concerning the shortcomings of life in India.

The call to higher things was roundly dismissed:

"Ashrams, Engleesh, they are mainly bullshit!" Sadhus were sadhus, "only because of the free food".

"They're probably not all rotten apples," I retorted, leaping a shade hastily to their defence.

"Why do you visit these holy places if you're so cynical?" She was beginning to irritate me.

This time an indifferent shrug, a grudging admission: "Perhaps I might see, meet someone – someone who will be able to tell me something."

"Such as?" I said quietly. She was calmer now.

I picked up a heavy negative vibe in the long silence which followed. Outwardly at least, she seemed determined to be unaffected by the legendary places she haunted. Ever the optimist, I tried the positive tack, casually alluding to the "energy" of the mountains. Again I was cut short: Only Tibet counted. I couldn't see much point in continuing such a conversation and lapsed into a silence of my own. This was no deterrent to the flood of opinion and after a further half-hour my head was so full of intrusive garrulity that I decided to act.

I got to my feet.

Still talking, although I had clearly stopped listening, the Italian did likewise, picking up my book as she did so. For a few moments I thought she had a mind to accompany me to my room and there continue the conversation or verbal rambling as it had become. But at the last moment she turned away still in possession of my book. I let her take it because she said she wanted to make a painting based on one of Govinda's illustrations of the Tibetan landscape. Some personal bitterness was boiling beneath that extrovert façade which had nothing to do with India, 'bullshit ashrams' or freeloading sadhus.

A short sigh of relief broke from me as I reached the sanctuary of my door. The Kashmiri's was firmly shut, but I could hear a breeze whirling behind the shutters.

No further melodramas disturbed the remainder of my first Sunday in Rishikesh. Around one-thirty I ate a banana. An hour later I heard the Kashmiri cough. When the declining sun finally began to peek round the edge of my window about four, I got up and took a shower, taking care not to swallow any of the water which ran down my face. When I lay down again I quickly noticed I had company, though this time of the harmless variety: an olive, splay-footed gheko which hadn't been there before was pasted to my ceiling. We exchanged stares in total stillness until I closed my eyes. When I re-opened them, it had moved three feet further away; still it eyed me from its upside-down position. I stared back, trying to catch it in motion. I never did. Strange games.

The hands of my watch crawled towards the infinity of early evening. The fan's whispering rhythm was cool on my skin and I remained put until about six, when a languid Western voice drifted by in the quadrangle. I got up and sauntered to the window. At last, life! A meandering group of foreigners was just filing past the broad temple steps twenty yards to the left of my room. I immediately stepped into my flip-flops as a dull metallic note summoned the hungry from their rooms.

A mingled aroma of cooking smells and fire-smoke greeted me as I ducked through the doorway of the kitchen-dining room, tucked inconspicuously away in the furthest corner of the courtyard. A few seconds passed before my eyes became accustomed to the dim interior. It was a low rectangular room with a stone floor, plain walls and squatting-sized tables along three sides. A murmur of voices wafted from the cooking area at the far end of the room.

Sitting amongst the stack of trays and blackened cooking pots an older man was quietly directing the same young helper from my neighbour's room the evening before. At his side was a middle-aged woman of serene countenance whom I later came to know as *Mataji*. The older man clearly had things under control, issuing occasional directives in Hindi to the others or raking at the fire with a metal rod. I was struck by how the convivial threesome's air of harmonious activity, of community, differed from our self-conscious individuality: fifteen or so foreigners, each motionless in his or her own space behind the low tables, waiting in our curious self-sufficient way.

Feeling like an actor attending an audition I sank into a semblance of the lotus position behind the nearest empty table. From my neighbour, a seasoned-looking traveller with a wispy beard and sun-bleached hair straggling over his shoulders, flickered an alert sideways glance; but that was all. I looked around at the others: we were a motley, variegated lot, I had to admit. Too many 90s nose studs and not enough naivety for the late 60s. But I caught a whiff of Woodstock, nevertheless. One of those fart-cushions might have cheered things up a notch. The doorway darkened again as several more Westerners, the Italian chatterer amongst them, sauntered into the shade, removed their shoes and padded towards the few remaining seats. There was no sign of the Kashmiri. A numbered lottery-type ticket was lying on my neighbour's table. I was about to remark upon it, when suddenly he stirred, swivelling his head to give me the full-on, unblinking stare which I had sometimes noted on the faces of yogis or the *chillom-babas*. He seemed about to deliver a gemstone of profundity, and I leant sideways to hear better.

"Nah worries," came the reply to my unspoken question, "you can pay tomorrow."

"Okay," I nodded, feeling a little let down at the content of the unadulterated Aussie twang.

The corners of his mouth twitched a fraction as if at some private amusement, before his gaze swung away from me to refocus on some indeterminate point in the middle of the dining room. The moment of telepathy was over.

It didn't need clairvoyant powers to guess what the meal would be. With busy, muscular speed and the customary grin the younger man scurried back and forth from the fire with plates. I was more than ready for my evening crust, however, and quickly emptied my tray. The tea was a different story: a thin, execrable brew without spices, a prime candidate for a Nixon-type denouncement. I took a few sips and left the rest. As I pushed the metal jug to one side, an ant popped over the edge of my tray and surveyed the leftovers. The meal was the price of a Mars bar back in England.

I took my turn at the cold-water sink adjacent to the far inner door, adding my washed pots to the glistening pile in front of the fire. When I told *Mataji* I would pay her the next day, though she seemed not to understand English, she gave me the feeling she had listened to such promises before.

"See you around," I said to the Australian, as I recovered my footwear and wandered outside into the languid, velvet evening.

I sat quietly on my own by the Ganges, listening to the rush of the current and wondering what it was which had drawn me to this particular place. Eventually I ambled back to my room. Tomorrow, they'd said, Swami Dharmananda was due to return after two week's absence. But however peaceful the scene by the Ganges, I had little appetite for hanging out indefinitely, and hoped his arrival might bring the purpose of Ved Nikatan and my own into sharper focus.

Despite the relatively early hour, reading soon left me heavy-lidded and I lay down in the semi-darkness, hoping this time I would have an undisturbed night. However, hardly had I extinguished my candle when I heard a faint but unmistakeable sound. Shrill, high and closing in rapidly – I raised an instinctive hand…

'Refrain from injury to all living beings', goes the Buddha's Precept for compassionate living. Hindus call it *ahimsa,* or harmlessness, the removal of the desire to kill. *Patanjali*, in his ancient text, *Yoga Sutras*, says, 'When a man becomes steadfast in his abstention from harming others, then all living creatures will cease to feel enmity in his presence.'

The peevish, circling whine grew steadily closer to my left ear…*Ahimsa*, it was a tall order in such circumstances.

Just how far had I journeyed on this road to noble virtue? The idea ran through my mind. I could understand it. Should I simply lie back and look on with loving kindness as the creature bit me? Or what? Even if, as I had been told, the district was a non-malarial area, I was still prey to the odd doubt. Dengue fever, meanwhile, a disease transmitted by infected mosquitoes, had broken out in Delhi in recent months – so there was no point in dropping all precautions. Suddenly I felt the insect's gossamer touch on my temple! In a split second, all my noble intentions vanished as – THWACK! I sent my tiny tormentor to infinity.

Is there a mosquito equivalent to the Buddhist's world of 'Not-self' or 'Not-mosquito'? I am sure it must be so. But the arguments cannot simply be slipped on a thread like so many beads; an additional chapter of somewhat dull exposition might well give the reader the right stimulus for an early night. But clearly – to return to my own action involving a creature smaller than myself – I still had some travelling to do. I was not yet a 'floating stone'.

Chapter 4

Dharma Talk

Dharmananda's unlined countenance nodded patiently at my question: "The experiences you have had during the first months of meditation – bright lights, cessation of breathing, great calm, and so on, are nothing unusual. They are the good *samskaras* coming up from a previous life; these thought forms are the signs of a very deep practice in an earlier existence. But because there have been no similar experiences since is not a thing to worry about – the purpose of this life is to carry it on. Even though nothing seems to be happening, development is still taking place inside us, whether we know it or not. In two years, five years – or perhaps ten, these experiences will occur again, but with even greater intensity."

I was sitting with the swami in the cluttered ashram office an hour after the evening meal on my third day. I had called to see him because I wanted to discuss one or two aspects of my own meditation practice, particularly on how to relate it to everyday life outside in the humdrum world. He was a tall, upright figure, glossy haired and black-bearded, and with an impressive presence. His face had a well-formed nose, a rather prominent set of teeth and an alert eye, suggesting both spiritual and intellectual vigour. I had the impression that few details escaped his notice around the ashram, also that he had a worldly understanding of people's ways and weaknesses. At a talk earlier in the day attended by some twenty five foreigners, he told of his own spiritual development. We sat or squatted on the floor in front of him.

Much against the wishes of his family, he had exchanged the uniform of an Indian army officer with high social status for the orange robes of a lowly Hindu monk. "The transition was a difficult one indeed", he remarked, with a wry smile.

There was also the feelings of his family to consider. His father, a high-ranking officer himself, was bitterly disappointed that his son should 'throw away' a promising career for a life as a penniless monk. After five years in the ashram, however, and now in his early forties, Dharmananda could reflect upon how such experiences inflict unnecessary unhappiness on ourselves when we have rigid expectations of others.

"My father was deeply unhappy man," he explained, "but slowly he is reconciled with the Path his son has chosen."

I listened quietly and with keen interest as he responded with gentle insistence to my promptings:

"There should be no fear or guilt about offending a few people because you will eventually benefit many more. Not talking – silence – can be of tremendous importance, yes. Family, friends – a few may be upset, irritated – say things about you. But this does not matter. Ghandi, even at the height of his political career used to maintain a silent period each day." Dharmananda's eyes laughed, "Imagine how this must have irritated people in government queuing up to see him with important business to discuss! No, for the ultimate benefits you should not be forced, by fear of criticism, giving offence and so on, to back against what you know is best for you."

Ashrams come in varying shades of spiritual practice and openness to outsiders. Some remain traditional, admitting no foreigners, whilst others, perhaps aware of the convergence of the paths, have adapted to the times and opened their doors to Westerners. Most visitors, having already accumulated an eclectic experience of both New Age and age-old, spiritual practices are not usually looking for a new religion, or a God who tells them how to behave or praises them when they do good deeds, but are actively looking for the Divinity inside their own selves. In a sense, these seekers are without a religion, pagan, almost. Some smoke dope and some don't. They see no incongruity in chanting Hindu chants and Buddhist sutras or doing Sufi dances and yoga, all in the same practice.

The teachers, too, are very different: the title 'Swami' or 'Yogi' or *baba* doesn't always mean that person is remarkably evolved spiritually. They, the gurus, are a varied species like everyone else. Some are very ascetic whilst others are more liberal. Some who have reached a high level of realisation don't talk or encourage disciples; many selfless individuals devote their time to serving the needy; a number succumb to the ego trip, abusing their position of prestige and knowledge, to the disappointment of more than a few of the travelling throng. The variation is huge.

Many Westerners actively seek out different teachers and stay at very different ashrams – "supermarket spirituality", if you like. I wouldn't care to say whether it is a wholly good or bad thing. Only that in the search for water, I am inclined to the view that one is more likely to find it by sinking one deep well than a host of shallow ones.

Whilst offering the possibility for the individual's complete integration of inner and outer life, an ashram, especially a bigger one, has to deal with the earthly question of how to fund itself. In the case of Ved Nikatan there was none of the unseemly opulence against which the doubting might vent their cynicism.

Funding for its various projects such as a library and visitor accommodation seemed to derive from donations as well as from room, meals and instruction charges. However, I heard Dharmananda, as the person charged with looking after the visitors' spiritual interests, more than once gently chide the management for their 'open-to-all' policy:

"If I could have my own way," he told one of the morning gatherings, "I would only admit those Westerners who have some spiritual purpose in coming here; too many use Ved Nikatan only as cheap hotel."

Internationally known ashrams, such as the Sri Aurobindo ashram in Tamil Nadu own considerable property and accommodate resident communities of several hundred people. Started by Aurobindo Ghosh in the early years of the 20th century, its subsequent expansion was guided by a Frenchwoman, Mirra Alfassa, known simply to her followers as *The Mother*.

Conceived as a kind of experimental city, its new centre opened at Auroville near Pondicherry in 1968 with a celebration attended by representatives of some 121 countries.

Though the 700 or so residents are actively pursuing a wide variety of projects such as handicrafts, education, health and agricultural research, the 30 years since its flourishing start has seen it develop into a not always harmonious creature, very unlike the *Mother's* dream of an international community of 50,000 souls. Internal squabbles and conflict with the local community at Auroville produced a great period of unrest culminating in the involvement of national governments. Only in the last few years has the dust begun to settle as arguments between the different factions were sorted out.

Very different in character, yet equally well-known amongst foreign spiritual seekers is Sai Baba's ashram at Puttaparti, north east of Bangalore in south central India. Naming it Prasanthi Nilayam, Abode of Eternal Peace, he teaches that the goal of human life is re-mergence with the Divine; our eternally unchanging True Nature lies hidden beneath accretions of worldly attachment to transitory experience. In order to achieve liberation (*moksha*) and discover divine bliss (*ananda*) we have to live according to the divine purpose without regard for the fruits of our actions. By seeing the Divine in all things, all actions, the seeker frees himself from ego, finds true wisdom (*Jnana*) and experiences God.

Unlike some spiritual masters, Sai Baba has never thought it necessary to travel abroad.

"Those who are drawn to see me, will come to India", he says. At his ashram and elsewhere in India he has performed miracles for which the sceptical have no explanation other than the traditional ones. He has manifested objects, resurrected the clinically dead, fed gatherings of many hundreds of devotees and made duplicate appearances in different places at the same time. Such activities stretch the boundaries of human credibility and are the insignia of an *avatar* or direct descendant from God. Jesus Christ would be regarded as another, though with one important difference: *avatars* are already Self-realised; they are born without *karma* and need no years of training to understand and show their limitless divinity. In this respect, such an incarnation is unlike a *guru* such as *Osho (Bagwan)* or the TM leader, *Maharishi Mahesh Yogi* who have all been through years of patient spiritual practice.

Two questions above all seem to occupy the minds of those visitors who leave Prasanthi Nilayam unconvinced by Baba as God manifest on Earth:

Why does he continue to materialise objects such as watches, rings, portraits of himself and so on, when such actions would seem to belong in the company of fakirs, magicians and other wonder workers of the occult? And why does he use his miraculous healing powers to cure some people and not others?

To the first question he answers quite simply that the tricks serve only to draw those who are interested towards him in the first place; people are always curious about yogis who have developed unusual powers. His chief concern is not vain self-glory on a mundane *personality* level, but to use his influence upon the soul-lives of these people. The second question is a more complicated one. Sai Baba reminds us of Christ's miracles and the fact that he, too, did not use his powers to cure all the sickness around him. A crucial aspect, according to Baba, was the individual's working out of his or her personal *Karma*: to what degree was a particular sickness *Karmic*? How far should a person's *karma* be tampered with? He saw his saving of some people from death as being no more than an outer miracle. The true miracles are the inner ones. When he has not saved the life of a dying person, he has followed that course with good reason. In place of an outer cure an inner healing has taken place; he has brought an acceptance of the wider implications of suffering and death... he has healed the desire to be healed.

It seems a wonderful thing that at the *Abode of Great Peace*, all one's spiritual uncertainties could be resolved forever, by a flashing glance of miraculous power from Sai Baba. Visitors report such transformations. Perhaps Sai Baba <u>can</u> transmit divine healing powers for which he claims not to be the source, but only the vehicle for their transmission. It would be better to visit Prasanthi Nilayam to experience for oneself before pronouncing with any authority on such mysterious happenings.

Ved Nikatan did not slot neatly into any of these categories.

Dharmananda clearly did not seek to project himself as a totally realised being, a spiritual master who could turn lost souls towards God with a single utterance:

"There is still some way to go," he informed us with a self-mocking chuckle, during one of the regular morning gatherings.

His arrival also brought a rapid and welcome improvement to my material circumstances. Amused by my account of nocturnal happenings whilst sleeping on the stone floor, he immediately instructed someone to remedy my needless austerity. With the provision of a bed, a Siva statuette which I bought in a bazaar, a few garish postcards about the walls and a handful of flowers from the forest – and my room began to shed its resemblance to a chamber of sensory deprivation. I was beginning to settle in.

A few doors along from mine was an extrovert German in his mid-thirties, a typical happy-go-lucky who had moved in about the same time as myself. He plucked amiably at his guitar as we exchanged notes on the sunlit courtyard steps one morning.

"One month ashram one month seaside when I am in India," he explained. "Next year I stay at home and my wife travels." His eyes danced as if reliving some Goa beach scene or in anticipation of seaside parties yet to come. Towards the end of my first week the Kashmiri disappeared. His door swung open around ten and another traveller of whom Swami Dharmananda might not have entirely approved sauntered into the sunshine.

A complete stranger to me, he ambled out to the edge of the steps without acknowledging my presence. A Westerner with black hair cascading about his shoulders, he wore a short cotton waistcoat dyed navy and faded cotton pants scissored off just below the knee. He squinted myopically into the light, hooked a *beedi* from its volcano-shaped packet and lit up. He inhaled, blew out a sigh of smoke and shuffled flat-footed back to his room. A few minutes later he returned carrying a plastic cup and sat down a few yards from me. He finished his drink in silence, put his cup down and gave me a quick once-over:

"Bob..Bin in Rishikesh long?"

'Bob', 'Bobby'? Could this be the generous, free-lending 'Bobby' on whom the hotelier had heaped such praise? He hardly fitted the bill was my immediate impression. We talked for a while in desultory fashion. Bob was also English with an all-too familiar reticence. Or perhaps he was just naturally quiet: it was difficult to tell.

"I think I was born in Watford," he said, after another fill-up back in his room.

I nodded as though that revelation explained everything. He spoke in a quiet, jerky way with long pauses for thought, during which my attention would start to wander. Before long, I was only hearing fragments and tail-ends of his 14 months around India. I began to feel the first stirrings of discomfort when he made a succession of uncharitable remarks about Indians. The tales of woe dragged on. In

65

Delhi he had been robbed of 300 dollars at knifepoint. Outside that same city his rucksack had been stolen on a train. He hadn't got on well with the Indian authorities, either. He was 'busted' by the drugs police in Goa and 'set-up' while staying in small hotels elsewhere. And so on.

"Cheerful times," I murmured to myself, his negativity beginning to lap at the edges of my own mood. I heard about Hampi in the south, where by 9.30 or 10 in the morning it was already too hot to walk barefoot on the stones.

I decided it was time to make a move before my new neighbour's jaded outlook on life took possession of me. Just because we were both from England didn't mean we had to be friends. I rose to go, reeling off some fictitious errand as an excuse.

Bob, unconcerned, thumbed his lighter for the umpteenth time. "Maybe we'll meet up in Manali – if that's where you <u>say</u> you're going?"

I couldn't recall saying that Manali <u>was</u> where I was heading, but I said, "Yeah, maybe," as I wandered off, pretending I was going for some diarrhea pills. Perhaps I had more in common with people like Bob than I cared to admit to myself; it's all too easy to project our own stuff on to others.

After breakfast on my fourth morning I made my way up to the room overlooking the Ganga where Dharmananda usually delivered his daily *dharma* talks. Two or three Westerners were already sitting cross-legged on the ancient carpet. I picked up a cushion and sat down towards the back. In India there is no escape from dust, and most of the statuettes and framed images of Hindu deities decorating the shrine-like room were much in need of a caring hand. When the orange-robed swami swept in a few minutes later, the assembly had swollen to around twenty.

He sat without moving a muscle, his eyes lightly closed.

After a brief prayer in Hindi, he drew breath and glanced with practised ease about the room:

"There are one or two newcomers here, today," he observed assuredly. "Mainly for their benefit, I will remind the rest of you of the purpose of these meetings."

My vaguely-formulated thoughts vanished as he began to outline what the ashram had to offer for the sincere visitor. Almost as if he was reading my mind, the swami's attention soon turned to a subject on which my own thoughts had dwelt in recent months, namely, the guru (or master) and his or her relationship with the disciple, or *chela*. A much overused term today, its original meaning has become cheapened in the West where many regard the notion of 'master' with deep suspicion, mistaking the *guru-chela* relationship as a personality cult. This is far from the truth. And the spawning of 'cooking gurus' and 'fashion gurus' is a misapplication of the word and a glib distortion of the guru's role in Eastern spiritual culture.

"How will we recognise a true guru amongst all the impostors when we meet him?" he asked. "Such a man must have certain striking qualities. He must have direct experience of God and not have merely read about His attributes in the scriptures, second-hand knowledge. This man will have profound qualities of humility and steadfastness; he will have reached a state of unblemished purity, a level of egolessness, a man free from all desires of the mind. Freed from attachment to people, places and things – he is free to Act. The mind of a guru will be steady under all circumstances. He will not need to travel about in search of followers – those who seek the truth will be naturally drawn to such people."

As someone sitting behind me cleared his throat, Dharmananda paused in mid-sentence, sensing an interruption.

"Why d'we need a guru"? asked a hesitant European voice.

The swami chose his words with care, speaking lightly, yet forcefully in a tone both easy on the ear and stimulating at the same time:

"The guru is the man who opens the eyes of the spiritually blind. But he is guide only, a man who knows the way to God. In the world, people have great difficulty in judging things in their true light, whereas the guru, living the holy life, is better able to judge the value of worldly goals, such as money , position or sense pleasures. If we want some legal advice we consult a lawyer – we don't ask the opinion of the man in the street, So if you are an earnest seeker you will need a guru at some stage. His knowledge is superior to all knowledge to be found in books."

His eyes roamed amongst us:

"Some seekers, not over-endowed with humility, scour the world in search of masters who are 'good enough' for their ego-fantasies. Such people have no real respect and will learn nothing from the guru."

He then spoke of hindrances in spiritual practice, his eyes constantly alert for signs of misunderstanding or puzzlement:

"Why do bad thoughts come to us? They come because there is something that corresponds inside. For example, think of the mental atmosphere of a public place – it is a tangle of thoughts, ideas and wishes. Your head, as well as your mind is in the middle of it. One must have a very steady sense of inner direction to live in harmony in such surroundings. A filter of consciousness operates; it makes you conscious of certain thoughts and not others. What governs this is your inner affinities, inner habits and attitudes. Therefore the nature of the thoughts you receive may be an important indication of the kind of character you have. An optimist will generally have optimistic thoughts. The more you want to progress, the more you must overcome all that is in you which doesn't want to progress.

"Feverish activity," he raised a forefinger in emphasis, "haste and agitation – it

67

is the great delusion to think that this changes things. It is like taking a stick and beating the water with it. The water is moved about, but it isn't changed for all your beating. Whatever has been achieved in the world has been done by the few who could stand back from their action in silence. It is ignorance to think you must chase about, labour from morning till night, work without rest at all sorts of activities in order to do something for the world. Once you can step back from these whirling forces, you see how great is the error.

Humanity," he smiled tolerantly, "is like a mass of blind creatures rushing about. This is what society calls action, life. In practice, living for the moment is not true, though from the psychological point of view it ought to be. If you are always hurrying, you are in a state of constant inner tension."

The air in the room was slowly heating up as the morning progressed, but there were no signs of inattention as the swami developed his theme:

"Many times we are hurt by what people say about us. We must learn to detach ourselves from blows given by ungenerous people, learn to look at such perversions of thought from a higher altitude, from where we can see them in their proper perspective – the impersonal one. When you feel weak when talking with people the cause is a weakened nervous force. You should take a rest when you feel these symptoms very strong and open yourself to the higher forces which watch over you and prepare your path. They can draw towards you things which help you, draw people, books, circumstances; all sorts of little coincidences which greet you as though brought by a benevolent will. These will give you support in taking the right decisions and turn you in the right direction."

As the two hours drew to a close, my senses began to respond to a mingling of sounds from outside. I heard a man's voice calling in Hindi, the slap of sandled feet along the corridor.

"On Thursday…" Dharmananda's voice pulled me back to the present – "we will be looking at Bhagavad-Gita. Any questions?" His gaze fixed on some invisible object for a few moments before he brought the meeting to a flourishing close with a short prayer.

The talk had proved more interesting than I anticipated and I felt that a few weeks stay at Ved Nikatan would feed more than a mere academic curiosity about life's big questions. Dharmananda had also mentioned some *pranic* breathing exercises and I was curious to know more.

Books had given me plenty of ideas but offered no real guidance in the practical experience of Yoga in its deepest sense. There were four hours of exercise daily; two hours in the cool of early morning at 6a.m and another two in the evening. I quickly decided that I preferred to perform a shortened morning session in my own room and called at the office to inform Dharmananda. He agreed immediately after listening to my reasons, and added that I needn't pay the full fee but should make a donation instead.

Over twenty people turned up for the first evening's yoga in the basement hall of the new temple. One or two Indian men also appeared, apparently viewing the somewhat unusual spectacle with idle curiosity, though I noticed on more than one occasion how their eyes strayed towards the female participants.

Dharmananda proved to be a no less vigorous teacher of yoga as he was in giving dharma talks. Dressed once more in spotless orange robes, he showed a commanding yet sympathetic awareness of the individual difficulties people might be experiencing.

He spelled out the aim of yoga:

"The overall purpose is spiritual growth, harmony, unity and to the Hindus – God. The yoga *asanas*, or bodily exercises, form the basic foundation of the Yoga Way. In all exercises you should try to breathe normally without overstretching. Don't treat them as gymnastics or let the face distort with pressure."

The next evening, however, I was surprised to discover a foreigner in his place.

A tall Englishman dressed in loose white clothing Indian-style, appeared at the front of the hall and introduced himself as "Peter". He informed the room that he would be taking occasional sessions when the swami was unavailable. Subsequently, Dharmananda explained that after a full day of his other ashram duties he would sometimes have too little energy to run the evening class. Peter, he assured us, was an excellent yoga teacher.

Just as the session began, a violent thunderstorm broke over the temple, an abrupt thunderclap coinciding with the *fizz* of a lightning flash. It was an eerie couple of hours, to exercise in that windy hall, whilst outside the storm crackled and blew with great fury. Later I had the opportunity to talk to Peter when I bumped into him in one of the bazaar's tiny cafes.

About the same age as myself, I soon realised that his reserved exterior concealed a very real passion for India and the spiritual path. Not a man to make conversation for conversation's sake, there were sometimes lengthy pauses as he pieced his story together: India had been his life for many years. Once, after badly injuring his back during yoga practice, he had not been home to London for seven years. Unable to walk, he was ferried about in a wooden push-cart or in a taxi, lying full length in the back. Because of his height – well over six feet – and the pain in his spine, he would sometimes resort to hanging his feet out of the window, a potentially lethal ploy if ever there was one. On reflection afterwards, I wondered if there was an element of exaggeration in what he told me – Why didn't he go home, for instance? But that was his story as he related it to me. Now recovered he would return to England once or twice a year – to visit his mother in London and to reacquaint himself with his roots. But the adjustment to the British way of life was proving a struggle. There were three stages of adjustment, he told me:

"One, it's fine for the first few weeks: you're carrying a lot of energy from India. Then there's the downward phase when you start to think of all you've left behind in India. Finally, there's the stage where you're no longer regretting having left India, but you're looking forward to the next visit. In the meantime you decide to adjust and make the best of being in England. It's good to go back to England for a while... the countryside, a little TV, a little sport. When I go for a walk in England, I often feel that this is the best place to be; it's good *karma* to be born in England. I can't think of any other nationality I'd like to be, not any Third World country, aspiring to the things of the West." He spoke quietly and with feeling about the late 60s and 70s in England and reminisced about his unfailing ability to find employment whenever he revisited his home town:

"I always made a good impression at interviews. If there were ten applicants for a job, I would invariably get it. But times have changed. For the past ten years there has been one big gap in my CV – employers are suspicious."

His voice faded away and he looked at me in expectation.

After I told him a little about my own life we sat in silence for several minutes, enjoying the ebb and flow of people passing the café entrance. When we parted, Peter towards the bridge, and myself back to the ashram, a ragged woman thrust out a basket of torpid snakes for our reluctant inspection. I kept my hand away from my pocket.

I attended the daily talks with a sense of anticipation for whatever themes the much-respected swami intended to discuss. It was by no means an unusual occurrence for the meetings to over-run their allocated span if some point required further clarification. One time I even spotted a *gheko* clinging motionless near the top of a nearby pillar, as if it, too, was intently listening along with the rest of us.

Days passed.

He spoke of the stages in spiritual life:

"As in physical development there are three stages, childhood, adolescence, maturity. Most spiritual aspirants, they fall into the second category. The *samskaras* (latent tendencies or thought waves of the mind) are compelling us, prodding us to go higher. Even if we go back to material enjoyments, we will find that we will not be able to enjoy them as before. It is a process – we should be prepared to feel the conflicting pulls up and down. If the idealistic things are not happening in the street, it doesn't matter. There is a conflict. But we must learn to stick to our spiritual journey."

"In the West," he added with a faint smile, "Yoga is seen as gymnastics, body development exercise. But in India, Hatha Yoga and all the rest are seen only as stepping stones on the spiritual path. Whilst each is not totally distinct from another, they are all regarded as subordinate to Raja (Royal) Yoga. The religious Yoga of a devout Hindu bears little comparison to the fitness and relaxation classes of the West. Its purpose is to unlock the spiritual powers which we all carry inside. The ultimate goal is *Moksha*, liberation. Yoga is the Path by which the aspirant's own spirit is unified with the Divine Spirit.

Raja Yoga has eight stages leading to God-consciousness. The most important of all is – *Ahimsa*, non-violence. This does not simply mean not-killing, but also the absence of violence in mind, speech and thought. Every weakness in the spiritual aspirant hurts him more than others because he is struggling to reach very high standards. When a man finally reaches the stage where he has no enmity, then nature outside him – animals and people, sense this and do him no harm."

"*Satya* (truthfulness)," Dharmananda's voice slowed for a moment as a latecomer entered the room, "that man who speaks only truth over long period of time – means whatever he speaks will happen. This is why people go to guru for advice because whatever the guru says tends to come true in other people's lives."

"*Asteya*, non-stealing, this also in the sense of not taking from society more than you need, as in food and clothing. *Brahmacharya*, control of senses, celibacy. This means moderation rather than celibacy in extreme way. If you forcefully abstain from sex without gradual preparation of the mind, then suffering will arise. As long as one feels the need it is all right in moderation."

Whatever the place's shortcomings, fate had dropped me at the doors of Ved Nikatan , and now I was there, I was much in favour of staying in one place for a while. I did not feel pulled to endless dusty journeys across the Indian continent on fruitless enquiry: *Shambala*, the fabled seat of spiritual inspiration, has not only a geographical location, but is to be found internally, in the heart as well – an intelligent reason for staying put, I thought. In some strange way I sensed that all I needed I already had. Not another book, another place, another talk, or more information was needed – just an inner shift in consciousness. If that was to happen at all, it could, beyond doubt, happen in Rishikesh as anywhere else in that vast land.

The sun had not yet risen above the nearby hills when I set out for a walk one morning before breakfast. I had no particular objective in mind, but I took to the stony lane which ran alongside the north wall of the ashram enclosure. Within a few minutes I entered an inclined area of scrubby undergrowth, trees and a bouldery stream bed which I used as a path. My step quickened as I tasted the air which was cool and invigorating at that time of morning. The stream bed soon

71

brought me to a narrow dirt roadway, where I turned left towards Laxman Jhula bridge, the road following the indentations of the steep hillside. After twenty minutes I noticed the vegetated entrance to a cave above a bridge which had a half-hearted stream trickling under it. I made a mental note of the place for future reference and carried on along the deserted road until my attention was arrested by a forested ridge rising to my right. High above, my gaze fixed upon a solitary tree which looked as if it might mark a good lookout point over the immediate neighbourhood. Spotting a faint trail starting up from the road and close to the foot of the ridge, I needed no further encouragement.

The path was little used and soon I was thankful for the assistance of a trailing branch or creeper as it finally faded out altogether amongst the close-standing trees. The forest's deep silence was broken only by the sound of my own breathing or a miniature rush of shale when a footstep collapsed. It quickly became difficult to keep a bearing on the lookout tree, and I continued upwards more in faith than certain knowledge of my position relative to it. I wondered what sort of animal life might inhabit these woods but I neither heard nor saw any sign to give me a clue. Even so I once or twice had the distinct impression that I was not alone. My sense of direction proved to be functioning well-enough when, half an hour later I came up to the tree which stood in a little clearing or step, where the ridge lost itself against the main part of the lofty hillside. I paused for breath only to discover that the spot offered no spectacular view of the Ganges, as I had hoped.

Continuing up the rim of a bush-choked gully, I suddenly came upon a well-marked trail. Scarcely had I wiped the sweat from my forehead when I received another surprise: one second I had the trail to myself – and in the next, a man was standing no more than ten yards from me. I had heard no sound.

He wore a grubby turban, was short in height and had thin, sinewy limbs suggestive of strength and much manual work. Unlike myself, he appeared not to have been taken unawares and was the first to break the silence:

"Where are you going?" he asked in English, glancing behind me, as though gauging whether I was alone or not. I was still regaining my breath and waved my hand vaguely at the hillside above us. It was then that I also noticed a twist of smoke curling above the tree tops a few hundred feet higher. At once the man's tone became more urgent:

"No – not that way! You go down footpath…very dangerous up there. Elephant!"

He made a series of threatening grabs at the air, as though to indicate what fate awaited me if I carried on. Seeing that I made no immediate move, he wagged his head, rolling his eyes as though in terror:

"Go down footpath!" he repeated, "elephant up there – very bad temper!"

I could see that the subject was not open for discussion.

I waved in farewell and began to make off downhill as if following his advice. Once out of sight round the nearest bend however, I sat down on a convenient boulder for a proper rest and to think things out for myself. After I had cooled down I came quickly to the obvious conclusion. The skyline over which the sun was already peeping was a further 800 feet of steepening forest and shaggy, beetling cliffs. Always assuming I would find a path to lead me there, the ridge might take me another two hours of thirsty ascent of unfamiliar terrain. I would be better leaving it for another time, particularly since I was disinclined to skip the swami's morning talk.

On the descent I considered whether the forester had been superstitiously cautious concerning wild animals. It was easy for one's imagination to start playing tricks, yet an occasional movement amongst the otherwise motionless leaves left the debate suspended in mid-air. Later I heard that there <u>were</u> elephants on the mountainside, not to mention a roaming tiger or two.

The exertions had awakened my appetite by the time I regained the still drowsy vicinity of Swarg Ashram bridge. *Choti Wallah's* was already open and I sat down at one of the freshly-swabbed stone tables in the centre of the hall-like dining space. Sweat trickled from under my hat as I glanced about for a waiter.

I wasn't the only person dressed informally. Standing nonchalantly in the lane barely a handshake from me was a tall Indian with hair of prodigious length. I had always understood from Western verities on the subject that men's hair in particular would not grow longer than about waist-length – but here was a man six-foot tall with hair down to his heels, with another two feet trailing in the dust! India never ceases to jerk the newcomer from his complacent assumptions of what is and what isn't the case.

Oblivious to the gaze of a spectator, he took hold of each of the five or six ropes of dark hair in turn and proceeded to wind them beehive-style about his head, the final result bearing a close resemblance to a turban of coiled snakes or a guardsman's busby.

After *Choti Wallah's* I wandered slowly back through the bazaar towards the ashram. The beggar-woman was already at work importuning the devout and not-so-devout as they trickled to and fro through the open gates of a temple garden. A glimmer of recognition crossed her face as I neared her and a mumbled "*Backsheesh!*" greeted me as I passed. But this time the basket lid remained closed.

In the *dharma-* room there was little floor space for late arrivals as the swami took his customary seat at the front. An open text lay before him on a folding book support. Speaking in clear and articulate English, he began a little more briskly than usual:

"*Bhagavad-Gita* – some of you may know – means *Song Sung by the Lord*. It was written about 3000 BC and is by far the most popular of the three main Hindu scriptures. The other two are the *Upanishads* and *Brahma Sutras*. The narrator is *Krishna*…listener is *Arjuna* – this man, he is representative of mankind. Gita's purpose: it is to resolve human conflict. It teaches many things. One, causes of life's problems. Two, their solutions. Mystery of life and death is explained to us. Duties in life – man is alienated from universe; from God. He may be independent of others but a slave to his own thoughts and ideas. He should go for higher sense of duty. *Bhagavad-Gita* gives us methods for gaining self-knowledge in these matters."

Swami raised his hand.

"Big mistake people make is to take truths in *Gita* as absolutes; but they are all relative. There are different paths for different people and for same people at different stages in their life's journey. What would be right for monk – no good for simple village man. We are all individuals with different needs. But – important point – before anyone can find Truth he must become the kind of person who can find it."

He gave a tolerant chuckle.

"It is obvious point – but people in the West do not know that the quality of a person's understanding depends upon what kind of life he leads."

The mental atmosphere of Dharmananda's talk was very different to what I had experienced in the West. It was as if he <u>began</u> with the Truth and worked outwards from the centre; the Western approach is to probe <u>inwards</u> from the circumference. Thus, the Oriental mind starts from <u>knowledge</u> of the Cause, while on the other hand, the Occidental sets out with a hypothesis and attempts to prove it. I did not listen uncritically but tried to offer no mental resistance. More than one Oriental teacher has commented that for Westerners, the biggest hindrance to spiritual progress is nearly always "intellectual arrogance". The head monk of a temple in Thailand wryly noted of his Western visitors:

"Nearly all of them – their cup is so overflowing with their own opinions, that there is no room for anyone else's. Until they learn to let go of their own views and opinions, they will learn very little."

Swami looked softly about the listening room.

"Religion should not mean dogma – 'Churchianity' – but (Self) realisation through a continuous struggle for mastery. It does not depend upon intellectual assent or dissent. It is not about learning but Being. If you want to be a Christian, you don't have to know where Christ was born or when he preached a particular

74

sermon. All you need is to feel the spirit of the scriptures. In India we have the greatest respect for the wisdom of your holy men, your saints. We regard Jesus as an *Avatar*, like Buddha…" His smile became a little wan, "Unfortunately, our attitude isn't often reciprocated by people in the West."

His talk regained its momentum and flowed on to embrace the Vedantic notion of the universe and its creation:

"The word 'creation' in English is in Sanskrit, <u>exactly</u> 'projection' because in India there is no sect which believes in creation as it is regarded in the West – <u>a something coming out of nothing</u>. In India we think of creation as a 'projection' of that which already existed. Whenever knowledge comes to unity from diversity that study is at an end. Thus Western science, which is approaching the conclusion that all is energy, all is movement… One, will come to an end.

All matter throughout the Universe, Hindus believe, arises from the one primal matter, *Akasha*; all forces such as gravitation arise from the one primal force, *prana*, acting on *Akasha*. At the beginning of a cycle, the latter is unmanifested. The outcome of pranic action is that more and more forms – plants, animals, men…planets and stars are born. The evolutionary process lasts a long time, but eventually it ceases and a period of <u>involution</u> begins. All is resolved back through finer and finer forms into the original *Akasha* and a new cycle begins. In the West," Dharmananda's tolerant smile re-emerged, "such pronouncements are regarded as unproven ideas. But we regard them as truth, being the discoveries of *rishis* in ancient times."

When he introduced the topic of *maya*, I relaxed, feeling I was on more familiar ground. I sometimes thought I had a little understanding on this subject, but as the Swami opened up I realised just how little I did know.

"Maya," he began, "is that which binds us to Time – Space – Causation. It is the great cosmic illusion which makes us think one thing when reality is something else. The West has misinterpreted the *Upanishads*. These scriptures do not say the 'the world is an illusion' – they say that it is we who are in a state of delusion about the world. Hindus say that 'all is *Maya*.' How to get above *Maya* is the task which the aspirant has to tackle. Man senses he has the power to go beyond *Maya*, and great conflict arises because of the desire to be both here and some other place."

In *The Law of Miracles*, a chapter in his 'Autobiography of a Yogi', *Paramahansa Yogananda* writes, "… To rise above the duality of creation and perceive the unity of the Creator was conceived (by the ancient prophets) as man's highest goal. Those who cling to the cosmic illusion must accept its essential law of polarity: flow and ebb, rise and fall, day and night, pleasure and pain, good and evil, birth and death. This cyclic pattern assumes a certain anguishing monotony.

After man has gone through a few thousand human births, he then begins to cast a hopeful eye beyond the compulsions of *Maya*… *Maya* or *avidya* can never be destroyed through intellectual conviction or analysis, but solely through attaining the interior state of *nirvikalpa samadhi*. The old Testament prophets, and seers of all lands and ages, spoke from that state of consciousness."

Dharmananda pronounced along similar lines: "*Samadhi,* is that state which the yogi reaches as his meditation deepens, a super – conscious experience in which the sense of 'I am meditating' disappears, is transcended. His consciousness has merged with the Creator; he knows the essence rather than the external form. In this state of cosmic consciousness he perceives the essence of all as light and he is now able to function outside of time and space. In the silence of *Samadhi* miraculous powers (*siddhis*) come to him, which he misuses at his peril."

My few days' stay at the ashram had now lengthened to a week. I had begun to recognise individuals in the bazaar and to be recognised by some of them. One afternoon, I spotted the snowy-maned Messiah sitting in impressive, rock-like pose in his customary place. Staring straight before him, he was clearly deeply immersed in abstruse, other-worldly matters. Perhaps this time I might walk on without being accosted. However, I was given no chance to speculate further for the man seemed to have a well-developed sixth sense for likely passers-by.

His eyes round upon me.

Is he about to deliver a Truth known only to adepts, some ineffable message from the depths, fraught with meaning for mankind?

His finger angles towards the firmament:

"ONE, SIR!"

Surely he has recognised an equal associate in concord and harmony? I flatter myself for the answer to this question quickly comes.

This time I meet the Aged One's eye and say, "Yes – but ONE what?"

My response seems to have touched a sensitive spot as, for the first time the Master's gaze drops; and in a second his power has evaporated. An apologetic look steals across his countenance as he answers half-ashamedly, "One cigarette, sir."

It is the last time he bestows grace upon me, and thereafter feigns not to see my passing.

I felt a tinge of regret for a while afterwards for having 'exposed' him and hurt his feelings. Perhaps it was my fault for having such unlikely expectations based upon appearances.

In that same bazaar I came upon another, a wandering ascetic seemingly of no fixed abode. However the lightness of his skin tone suggested a birthplace far from Oriental shores. He was standing in a long line of sadhus patiently shuffling towards a free meal set-up by the riverside *ghats*. Greying hair fell to his shoulders and a frayed loincloth, which made me think of Tarzan, was his sole concession to public decency. He looked about sixty, had a strong profile and stood head and shoulders above most of those around him. Some quality – perhaps his abstracted air – made me curious to know his country of origin.

On my way to visit the famous Sivananda's ashram I came upon him again. This time he was leaning over the parapet of the suspension bridge, about halfway across. I slowed impulsively, feeling an urge to make his acquaintance. He was gazing downriver and seemed too insulated from his surroundings to respond to a normal greeting. I could feel no current projecting outwards to the world of tangible objects and other people, so I stopped as if by chance and gazed idly down at the torpedo-like fish. After a minute I mastered my own feelings of reticence, of not wishing to intrude upon another's solitude. The vista of sparkling turquoise water and forested hills was sublimely peaceful.

At first I thought he hadn't heard me. Then his face, darkly etched by long exposure to the Eastern sun and the vicissitudes of an uncertain life (unless I was very much mistaken), half-swung towards me. His greyish eyes gleamed inwardly for a second, but I had the impression that my presence as a flesh and blood being didn't really register. His voice, when he spoke, took me totally by surprise:

"I've been here five years actually," came the reply in distinctly southern counties. Decidedly incongruous considering his jungly appearance.

I pondered this hard-won confession before venturing the obvious:

"You're from England, too?" I knew as soon as I opened my mouth the question was wasted. A well-formed hand with tapering finger ends flicked the air as if brushing away an imaginary fly. My query hung suspended, unanswered: ridiculous. At length, in a hollow voice as though he addressed himself, he offered an answer of sorts:

"I'm sixty two. I used to be from Devon."

He spoke the name 'Devon' as if it was another galaxy a thousand light years from Earth. His eyes warily scanned the ochre and white bank of the *Ganga*:

"They're supposed to be sending out my pension but..." His palm lifted airily as his voice tailed off.

I looked in the direction of his gaze, half-expecting to spot a stubby ER mail van scuttling towards us. But there was nothing. A nullity of heat haze shimmered above a few nondescript buildings which marked the line of the road. There was nothing to see.

Despite his obvious signs of former physical vigour, I sensed a vulnerability in 'Greystoke', a reclusive desire for privacy, to be left alone to go his own way. As his eyes drifted back to mid-river I glimpsed a strange look which might have been a polite warning that to probe further would be unwelcome. I don't think he noticed when I moved on.

We passed each other twice more during the succeeding days. On each occasion he was strolling alone through the bazaar, mumbling an incoherent string of Hindu syllables; and the same unconscious swatting of the air by his cheek. No sign of recognition crossed his face at my greeting. Whether his inner solitude was a sign of mysterious gifts, I could not say. Perhaps he had flipped mentally and couldn't get back? When I didn't see him anymore I was a bit anxious in case he was lying up sick in a cockroachy room somewhere. My positive side liked to think he had returned to the jungle.

Seated cross-legged on his low cushion, Swami Dharmananda sighed as he gently chided the foreign gathering:

"If you have come here with spiritual purpose, try not to get involved with too much social chit-chat with other Westerners. Much energy is wasted in talking. You should have periods of quiet every day; by doing so, you cut out an important source of distraction and the energy you save gets transmuted into spiritual energy. When you have to speak, remember to consider what effects your words will have on the feelings of listeners. Overmuch talking is a bad habit; it keeps the mind always outgoing. So don't sit around *Choti Walla* drinking tea and making lots of friends…"

Somebody tittered.

Dharmananda raised himself higher, his eyes pretending disbelief.

"There is a man in this ashram – he drinks 15 cups of tea every day. It is too much! You should watch what you put into your stomach as well as into the mind."

He cautioned us about non-attachment to the vanity of obtaining mysterious powers through spiritual practice.

"Only last year a Westerner was here, an Englishman. He said, 'Swami, I am only here for two weeks; can you give me a mantra? I want to walk on the water.'"

The Swami's teeth shone luminously as his face split with a smile.

"There is a story to illustrate the folly of such desires:

A son went to his father and said, 'Father, I have decided to leave home to seek enlightenment; I have come for your blessing.' His father acceded straightaway. 'Go, son, find a spiritual master and in eleven years come back and tell me what you have learned.'

Eleven years later the son returned home much excited.

'Well,' said the father, as they were standing by the river. 'What have you learned after all this time?'

'Father,' said the son, 'I have learned to walk on the water.' Thinking to impress his father, the son walked across the river without getting his feet wet.

78

However, when he returned, his father did not react as he had hoped. The father called to a nearby boatman, saying,

'Row me across to the other bank.' When he returned, he dropped three paise into the man's hand. Then he turned to his son.

'You have been away eleven years, and in all that time all you have learned is how to walk on the water! I can cross this same river without getting my feet wet for three paise – what you have learned in eleven years is not even worth one rupee! Go away and learn something of value.'

At that moment, the scales fell from the son's eyes and he realised the Truth.

'Father, I have been away for so long, but it is only now that I see the vanity of what I have learned.'"

There are those who actively seek out freakish experiences; others, through no fault of their own, have them thrust upon them, as illustrated below.

The unassuming face of another Englishman, perhaps 15 years Bob's junior, flickers briefly into focus in the gallery of ghostly faces which sometimes fills my mind's eye. His name was John, an art history student in his early twenties.

Seeking to pursue his studies of the Persian influence on Indian art, he took the night train from Delhi to Lucknow in northern India, the train trundling him across the plains at a typical forty miles an hour. A man sitting alone nearby looked at him strangely from time to time, though the few other passengers took no notice of the foreigner. There seemed to be no cause for alarm as the train clattered through the darkness.

I listened in silence as he told me what happened next.

Without the slightest warning the man began to shout angrily at him in a torrent of Hindi and occasional English words. Why he a total stranger, should be the target of such abuse, the student had no idea; the Indian was obviously very disturbed about something and kept up his non-stop verbal attack for several minutes. His manner became increasingly aggressive, giving the young Englishman cause to fear that he might be assaulted physically at any second. By this time the other passengers had begun to take notice. They told the man to be quiet, that he shouldn't treat visitors to their country in such a way. Couldn't he see that the foreigner was tired?

The man paid no heed and continued his rantings unchecked.

At this, several of the passengers lost patience. Rising from their seats as one, they rushed upon him, flung open the carriage door of the moving train, and pitched him headfirst into the night.

After a pause I asked what the man might have been raving about.

The student hesitated, "He was saying something about India – pre-Independence, but I wasn't sure what…"

An unpleasant event, nevertheless, which gave a strange feeling.

Tales of baffling powers and mysterious phenomena invariably excite Western curiosity, giving rise either to an out-and-out scepticism of the subject or a half apologetic admission that "there might be something in it".

If there is one Asian fakir's trick which has lodged forever in the Western consciousness – it is the Indian rope trick. I met a man in Rishikesh, a New Zealander, who claimed to have witnessed this extraordinary spectacle.

This is what happened.

He and a companion had made an overnight stop in a small town in the state of Bihar, a place not much frequented by Westerners. While strolling about the town in the evening, their attention was drawn to a throng surrounding a young boy and a much older man. As they joined the perimeter of the group, the man produced a length of thick rope and, with a well-practiced flourish, flicked it out to its full extent along the ground. The boy, meanwhile, was squatting in the dust, apparently taking no part in the proceedings.

The man had a fiery presence and a mesmerising gaze, which totally held the attention of the expectant gathering.

"I could hardly pull my eyes away from the guy's face," the New Zealander said, "it was a blindingly hot evening, I felt a bit strange…" His mouth twitched as he neared the high point of his story:

"All of a sudden, the rope was standing vertically in the air – we both saw it… a good ten feet high. The guy was talking all the time, very fast. He said something to the kid. The boy got up, walked to the rope. Then – I saw with my own eyes – he climbed the rope, the full length, as quick as you like. When he reached the top, the man shouted an' he came down again. I couldn't believe what I'd just seen, I tell you!"

I shook my head wonderingly. But I had not yet heard the story's equally sensational conclusion.

The man's companion had shown sufficient presence of mind to take a photograph of the boy climbing the rope.

When the film was developed they were even more astonished. The print showed the boy 'climbing' the rope as the two men had witnessed, but instead of rising vertically, the rope quite obviously had not moved from where the fakir had first placed it. According to the evidence of the photograph, the boy had never left the ground! The New Zealander regarded me with knitted brow, as if expecting an instant rebuttal of the whole extraordinary event.

I shrugged. "I don't know how you can easily verify such a thing. Some of those people, perhaps through –"

"Yeah, hypnosis!" he completed my sentence. They're able to make us see what they want us to see, I dunno."

In the face of the inexplicable we both soon reverted to silence.

So far on my journey I had not encountered anything quite so mind-boggling as that reported by my acquaintance from the southern hemisphere. The only time I had been sucked into a street crowd was in Almora, when I quickly discovered that the focus of attention was not a magician – but a television set!

I must admit to a grain of sympathy for these characters – after all, they are only trying to make a living, even if they do take advantage of people's insatiable desire to believe in unusual psychic forces. Yet I have heard more than one 'modern' Indian roundly dismiss such practitioners, for "they fulfil no economic function, are charlatans to a man and prey unashamedly upon the gullible and simple-minded".

Just what the besmeared ascetic hopes to gain from holding an emaciated arm overhead until his nails become talons over six inches in length – it is not easy to decide. Likewise the man who draws a curious audience by thrusting sharpened instruments through his cheeks with apparent indifference. Such public spectacles of self-imposed suffering may well be rewarded by a few coins in the bazaar, but they have little in common with the kind of life led by the true sages.

The problem is, where to find them? As I have previously mentioned, many of the true holy men have begun to find the modern Haridwars and Rishikeshes less congenial than of old for the undistracted pursuit of spiritual goals. Many have removed themselves to remoter regions where few will disturb them.

Thus the average Westerner, invariably pressed for time when he visits these 'homes of the seers' with the hope of meeting some transcendental being of infinite wisdom, may well leave disappointed. If he is unfortunate to encounter only a gruff sadhu or two, or some learned *pandit* who browbeats him with parroted scriptures, he will have good grounds for feeling he has been deceived. Understandably, he has no wish to be taken for a fool and has a ready-made excuse for accepting the prevailing scepticism. Contrary to his expectations he heads for home less of a philosopher than when he set out.

But just as there will always be verse-mongers amongst true poets, who cater for the popular public taste, so it is unfair to tar the minority with the same brush.

It is not generally realised that long before they became tabloid headlines in the 1960s and even before Ghandi's visit to Britain in 1931, several of India's spiritually advanced beings had already carried their testimonies to the West in person. One of these rare men with whom the reader may be familiar is Sri Paramahansa Yogananda, whose reputation continues to grow with the passage of time. His book, *Autobiography of a Yogi*, has been translated into almost twenty languages. Born in 1893, in Gorakhpur close to the Himalayas in north east India, Yogananda had a remarkable childhood and was clearly destined for an out of the ordinary life. He became a disciple of the renowned *Jnanavatar* (incarnation of Wisdom) Swami Sri Yukteswar Giri, who introduced him to the ancient science of *Kriya* Yoga. Commended many centuries before by Patanjali in his *Yoga Sutras*, *Kriya* is both a way of meditation and an art of living, which brings the soul to a union with God.

His mission to the West began in 1920 as a delegate to an international religious conference in Boston. For the next ten years he travelled widely, lecturing in almost every major city, sometimes speaking to packed halls of 3000 with many turned away. But he was far from being a kind of Hindu Billy Graham, seeking to convert his audiences to a narrow body of religions dogma. He strove to unite East and West, pointing out that Truth is the possession of no one religion, but is the foundation which unites them all. Regarded as a world teacher who brought the art of spiritual living down from the lofty heights of abstract principles to the world of personal experience, he was a believer in unity through diversity. His teachings did not promise the *beyond*, but the *here and now*. There was no need to 'become an Indian' in order to profit from his instruction in *Kriya* Yoga. He was fond of illuminating his addresses by citing instances of personal experience during his years in America. On one occasion, he told how shortly after giving a talk in New York, a man approached him brandishing a gun. In his *Autobiography* he described what happened next:

"'Do you know I can shoot you?' the man said.

"'Why?' I asked calmly. My mind was on God.

"'You talk about democracy.'

"He was obviously a mentally disturbed person. We stood in silence for a while, and then he said. 'Forgive me. You have taken away my evil.' He ran down the street as swiftly as a stag."

Dharmananda was never slow to confront us with our Western preconceptions, either.

One morning he suddenly announced:

"You should not consider Siva a Hindu god only!"

A mischievous smile played about his lips as he registered the collective air of bemusement.

"You should not think this is the case," he reiterated, then paused as though withholding the key point.

"Some worship as if Siva is literally like that. No, the form is simply created to represent, to symbolise, the Ultimate Power of Principle – God. People cannot worship an abstract principle – the image is simply a method which aims to bring noble virtues and powers of Good into the devotees' minds. Siva represents that which is auspicious, eternal. Accordingly, Siva need not be worshipped only by Hindus.

He flung out an arm towards a knot of curious bystanders at the hall entrance.

"You have probably seen them – Siva devotees – wandering about the streets and in the Himalayas. They smear themselves with ash, carry tridents and walk around in next to nothing – huh?"

One or two heads nodded in agreement.

"If they looked like that in the West, they would be locked up!"

The hall rang with a welcome burst of laughter.

The swami's dark face was serious once more.

"It is important to understand the symbolism. The ash is not to frighten people. It signifies the renunciation of perishable objects in the world. On a practical level, *pasma*, cow dung ash, is a very good disinfectant. It also gives warmth when smeared on the body in cold climates. Siva is called 'the *Destroyer*' but this does not mean that followers have a love of destruction. They worship out of respect for the knowledge that all things must come to an end, and from that end will come a new beginning. Ignorance as well as 'all things' will be destroyed'."

Dharmananda raised an orange sleeve to indicate the sentinel statue on the platform behind him:

"Siva dancing very happy....'Lord of the Dance' – many Hindu gods are known by different names. The rhythm of the dance – it controls the energies of the universe. Left hand holds fire, symbolic of destruction of all that lives. Right foot crushes dwarf, signifying ignorance of Man."

His finger traced a shape in the air:

"Circle of flame represents cycle of Time, no beginning, no end. Right hand raised indicates..."

...But my attention had temporarily drifted away from symbolic things, back to my rented stone cottage on the edge of England's northern Pennines. Though I lived simply enough by Western standards, compared to Dharmananda's *Shiavite* ascetics could I deny being an out-and-out materialist? Was I merely deceiving myself, thinking I had 'risen above' the herd instinct to surround itself with comforting objects, merely by walking away from them?

"Of course, there is no need to concern yourselves too much with symbolism." Dharmananda's charismatic voice swiftly recalled my mind from its ramblings, "What you learn in this place, whatever you read in the *Gita* – you can apply to situations in everyday life. God, you should understand," he veered off at a tangent, "is not a Being who lords it over the universe – He <u>is</u> the universe. Creation, it is part of His very essence. In the West, you usually think of God as Father, but here in India we consider God to be both Father *and* Mother. Wisdom is the Father. Nature, with her bright stars, streams, birds, animals, flowers, mountains – the boundless world of Creation – She is the Divine Mother. Jesus spoke of God as Father. There are saints who speak of Him as Mother. He is -on the cosmic level – neither Father or Mother. On the level of human relationships, His feeling is manifested in the woman. Reason governs the father's nature until he develops that aspect which never errs – intuition."

His eyes flashed.

"It would be a good idea if men thought of all women as mothers if they are on the spiritual path – it would solve many problems! But," He glanced about the hall with a kindly laugh, "the women in this ashram would not be too flattered to have handsome Westerners addressing them as 'mother'!"

"Even today," he returned to his main theme, "many Westerners erroneously think Hindus are backward people, that we worship idols. This is not the case. All the images you see in the homes and in the temples – their purpose is to focus the attention on some particular quality of the Supreme Being. This is why the gods sometimes have different names. It is a way of revealing different parts of their characters. Sometimes, for example, *Parvati*, mother of elephant god, *Ganesha* and wife of *Siva*, is shown riding into battle on the back of a lion; her appearance as Durga – waving ten arms is fantastic, but she is not revered out of fear, Her ten hands indicate a fruitful life; she is at work both in the heavens and down below, with earthly concerns. Her son *Ganesha* implies a fruitful outcome from a reverent regard for Mother Nature. When she is shown as *Kali*, wearing a garland of human heads she announces both the terrible and benign aspects of nature."

I had felt uneasy with such images. Her jet black appearance, blood trickling from her tongue, and a belt of severed human hands certainly evoked terror – but gentleness? It was difficult to perceive nobility in such a ghastly image.

Dharmananda shed further light on the subject:

"The point of the skulls is the inescapable truth that time consumes all beings; whether they are just born, the child, mature man or old-age pensioner – in the end, time devours them all. Mother Kali is black because she is personification of past and future which are both shrouded in the unknown, darkness. The timid are afraid of *Kali*. But the fearless and persevering, She leads back to the *Brahmin* state, the Supreme Reality. *Kali* followers are sometimes worshipping in graveyards or at the cremation grounds – very macabre to non-Hindus! There are some sects it is true, which have degenerated into corrupt practices, but meditation at the cremation ground can give deep understanding of the transitory nature of the phenomenal life. The point is realised that the universe, *Jagat*, is in a constant state of flux"… The Swami permitted himself a short chuckle. "The entire embodied world is in constant motion – as your Western science has begun to discover."

"Dis means that East and West – dey come together?"

Dharmananda's eyes picked out the questioner, a quiet Belgian in his early thirties.

"Thousands of years ago, the *rishis* and yogis were already exploring these questions; the nature of the universe was revealed to them through many years of profound meditation. The Western way has been to examine the material world, what is observable to the senses – and to work inwards towards the centre. Now East and West are converging in this respect. There is a book by an English scientist, Stephen Hawking – you probably know it already: *A Brief History of Time*. Best seller in this country also."

I spoke involuntary at this.

"In the light of what you've been saying, do we need to read it to *know?*"

Dharmananda paused before answering:

"You cannot simply throw off your Western ways like a sack of potatoes – books, particularly in the beginning can be a great help. Holy books however, are only signposts; they only point the way to God. But reading about God is not the same thing as experiencing God. People are often very proud of their book-learning, their thoughts borrowed from the wisdom of others. Our *pandits* will talk of Brahman, of philosophy, of the Absolute and all the rest. But scarcely any of them have realised what they are talking about. Most of them are good for nothing! To explain God after merely reading and thinking about the scriptures is like trying to explain to someone the city of Calcutta after seeing it on a map. Merely to read many virtuous sayings in the scriptures is not enough: the virtues have to be practised in one's life. So long as the bee has not tasted the nectar of the flower, it hovers around humming. Likewise, the nearer you draw to God the less you are inclined to question and reason."

I was reminded of my days as a philosophy student: learned discussions about the existence of a piece of wax, the tutor who warned us about 'putting Descartes before the horse', and 'utilitarianism as a normative system' – it was all so intoxicating at the time. Now it seemed as far from my present experience as Devon was to 'Greystoke'.

We move on. The infinite cannot be communicated by the intellect alone. Proving that God exists is of little interest to the Indian mystic. He reverses the whole thing and says, "What *exists* is God."

Contrary to popular Western belief, the notion of one God is the same in Hinduism as in Christianity. In India different sects have developed a particular affinity with one or another god or goddess; the statues raised to them do no more than express ideas about God. For the Hindu, worship of an infinitude of 'little' gods is only a path of stepping stones to an understanding of *Brahman*, the one great God.

Very important to Hinduism concerning the nature of the *One* is the notion of the *Avatara* (Incarnation). There are many legends of Gods coming down to the world for diverse purposes, sometimes in the form of a boar, a tortoise, a dwarf and a host of others. *Ramakrishna* is very clear on this often misunderstood point:

"The *Avataras* are to *Brahman* what waves are to the ocean. The *Avatara* is always one and the same. Having plunged into the ocean of life, the one God rises up at one point and is known as *Krishna*, and when after another plunge, He rises up at another point, He is known as *Christ*."

When the *Avatara* come into the world in this way, to a lower plane of consciousness, only a few in this world are able to recognise that behind the temporary screen is a divine reality. Why is it that such prophets are frequently not honoured by those closest to them?

Again *Ramakrishna* colourfully explains:

"There is always a shadow under the lamp, while its light illumines the surrounding objects. So men in the immediate proximity of a prophet do not understand him. Those who live far off are charmed by his spiritual glow and his extraordinary power."

These divinities my be understood both as historical and allegorical, persons. Jesus, the earthly manifestation of Christ consciousness, is widely revered in India, alongside Krishna, as one of the great *Avatars* – both have profoundly influenced the whole world. Their words in the *Bhagavad-Gita* on the one hand, and in the New Testament on the other, are both books of supreme spiritual guidance. Whilst it would be unwise to attempt a simplistic comparison of them as teachers, yet there are interesting parallels in their personal lives, as much as we know of them. Moreover, need there be any conflict in the notion that truth may be sought down a variety of paths?

The evening began with hints of good fortune to come. When I sat in silence in the cool shade of my room no distracting thoughts and sensations diverted me in nearly three hours. I forgot my bodily existence and felt I was about to float towards the ceiling; light as gossamer, my breath slowed until barely a trace of air entered my nostrils. A wave of bliss lapped through me and I felt surcharged in every part by a purer kind of energy. A procession of ancient, sage-like beings floated across the screen of my mind, dazzling me with their marvellous forms and radiant expressions. It seemed I was at the gateway to superior things. But close on the heels of a calm evening comes night.

Chapter 5

A Jungle Walk

I awoke several hours later, a queasy sensation in my stomach and the feeling of having been tossed around all night in a rough sea. When I sat up I felt faint, whilst any sudden movement caused the room to revolve like a fairground ride out of control. After a few minutes the nausea grew worse. Finally, there was only one place to go…When I returned some half an hour later I was able to down a glass of mineral water, before flopping on to my bed, where I lay like a corpse until long after the untuneful gong had clanged the signal for breakfast.

It is easy to imagine the worst on such occasions, especially if one has turned the fever-hung pages of those medical guidebooks concerned with tropical diseases. But around mid-morning I felt recovered enough to mix a rehydration drink of glucose, bicarbonate of soda and a little salt. A shower acted as a real tonic and afterwards I reclined on my bed waiting patiently as the unpleasant sensations slowly subsided. A corner cobweb fluttered in the fan's waft and once a monkey, bent on some pilfering quest around the ashram, bounded past my grimy, sunlit window. Sometimes, when my gaze focused upon particular spots on the floor or configurations of light and shade on the cracked walls, patterns and flowing, fantastical landscapes would begin to materialise, as though from some clairvoyant dream. There were also faces. I saw long mystical visages with beards and all-seeing eyes; others were less attractive with mournful, long-suffering expressions. One or two were nightmarish terrors which hovered on the brink of bestiality. Before long faces began to jump out at me wherever I turned my eyes. Only when I heard a key grating in a lock somewhere down the corridor did my perceptions begin to return to normal and the fresco of grotesques fade from view as the walls once more became walls.

I peered out and saw Sanjay, one of the ashram's monks, showing two new arrivals into a vacant room. Californians, I later discovered. I heard a brief exchange between them, followed by the soft slap of lightly-clad feet padding in my direction. Sanjay's moustached face hovered quizzically in my half-open doorway.

"You are sick?"

I nodded, "A little."

Probably used to keeping an eye on the health of foreign guests, his glance swept my room:

"You must take complete rest, a little water and no going for long walks. Later you can take small soup; bananas, too, they are very good."

He inspected me for a moment longer then bobbed out, "Look after yourself."

I was touched by his consideration, especially as we had exchanged no more than a nominal greeting or two since my arrival. As Sanjay's steps faded, I reached for my small cracked mirror to check what evidence had prompted such an instant diagnosis. My room was too dim for me to see my image clearly, but the good point was that after a week in India my face had lost its English pallor. My eyes, though, looked bloodshot, not a pretty sight. Somebody had told me that a Self-realised master may be recognised by a number of distinguishing signs in his personal appearance, one of them being – red eyes. In my own case of course –

I waited until the middle of the afternoon had passed before venturing from the shade of my room. The sun flashed in my eyes like a Sikh's *kirpan* as soon as I set foot in the still-dazzling courtyard. I headed across without delay. Halfway over, a man's arm suddenly lifted in greeting from beneath a roofed canopy standing on tangerine pillars to my left. My thoughts were elsewhere, but I halted and started towards him. The prone figure pulled itself to a sitting position as I approached:

"Hi, how's it going?" came an easy Canadian drawl.

"I was sick," I replied, mounting the two steps to join him on the stone platform, "but I'll be okay in a day or so."

By the Canadian's side was a jar of peanut butter, the remains of a bread-loaf and a wilderness-type clasp knife. With his all-American haircut, chiselled masculine features and powerful physique, he wouldn't have looked out of place staring form the cover of *Sports Illustrated*.

A mordant grin opened across his tanned face:

"I noticed you weren't around at breakfast." The grin broadened. "I figured you were either in your room <u>dead</u>, or you'd gone for a hike in the mountains – I was gonna check your room in a coupla' days!"

"Thanks for the concern," I said, with mock gratitude.

The first I'd known of his existence was two evenings before when I was sitting in the back of the tin restaurant. Someone at my table murmured, "There's Marshall," when a blanket-wrapped figure strolled by the entrance without seeing us. Twenty-nine or a young forty – I'm not very good at ages – he never told me his name in the three or four weeks we were acquainted. I don't remember telling him mine, either, when I think about it. 'Marshall' <u>might</u> have been his name – I don't know for sure. Names can be just more baggage when you're travelling, seemed to be his style. I gleaned a few facts about his life when we weren't being laconic with one another:

88

Both his parents died young, his mother within the previous twelve months. A brother lived in Australia and was expecting a visit. His profession he described as, "looking after a holiday island on a big lake in Manitoba". He had a Master's degree in something or other. I told him it sounded like a dream job – if the blackfly left you alone in the summer months. Noticing my glance, he stabbed a finger at the food, "Sure you won't try a slice a' bread an' peanut butter?"

I shook my head, "I can eat that kind of thing at home every day if I want – when I'm in India I try to keep off Western –"

"Yeah, do as the Romans do," the Canadian interjected. "You're right; I just felt I wasn't getting enough of the proper foods, here. A yogi bakes them all an' sells them in the bazaar."

"My energy levels are up and down all the time, at the moment," I responded, though initially I hadn't wanted to get too well acquainted. It's wrong I know, but the insensitive streak in me had quickly affixed the tag, *Athlete...Pumper of Iron*. But except in airport romance novels, people are rarely so easily pigeon-holed. The Manitoban's face moved as if some inner struggle was coming to a head:

"Y'know," he dragged the words out, "I'm really having a hard time filling my day in this place. I get up. I do yoga. Take a shower. Eat breakfast. Write a few letters...read. Then my brain goes! I spend mosta the afternoon here, or..." He gestured towards the open door of his nearby room, "on my bed."

"I used to be like that," I nodded, "but not so much now...What about the talks?" I had never seen him in the shrine room and wondered why.

"I went to one, but I'm not as interested in the lectures as most people here."

"They aren't like that...narrow dogma. He isn't trying to convert anyone to Hinduism...If it was just another culture's theology, I wouldn't be there."

The Canadian was unimpressed:

"I've been reading *The Celestine Prophecy*. It's an amazing book – every single incident in your life has meaning...part of the big picture."

I already knew that. The writer was just dressing up old truths in the latest jargon... Life isn't that complicated. Some of the wisest people are simply here and now. They tell you much by speaking little. Being here and now – easy to say; very, very difficult to realise. So long as our minds take us from the here and now, duality remains...we aren't part of the whole in an aware way. But at least the planet's heading in that direction. Unless you're a pessimist.

I let him finish before changing the subject.

"Tomorrow, or the day after, I'm heading up there for the day." I pointed out the route I would take to the forested ridge which filled the skyline behind Ved Nikatan.

"There's a temple called Neelkanth; I heard it takes four to five hours up."

89

"Yeah?" The Canadian looked interested and ran his fingers over the golden stubble on his chin. "Maybe we can do the climb together – I get such a build up of excess energy, here. I guess I'm just not used to so much inactivity."

I said, "I need to go alone…maybe next week the two of us..?"

He might have been disappointed but didn't show it:

"Sure, that's fine. Let me know when."

Two mornings later I felt well enough to leave. I filled my wineskin at the quadrangle pipe, took a quick breakfast in the sleepy bazaar and set out for the lane I had used on my first walk. The lower end was clogged with still-recumbent cows; a stray dog nosing for scraps in the gutter looked at me hopefully. I snapped the tip off a bread stick and tossed it to the hungry creature, which took it in a single gulp, staring wistfully after me as I threaded my way amongst the dozing cows. I took a bite of the dry stick myself, storing the remainder in my cotton shoulder bag for later. Bars of pale sunlight made zebra stripes across the dirt road of my earlier foray, which I reached within ten minutes. When I paused for a moment to glance back from the road's edge, a burnished temple roof by the Ganga gleamed up at me through the thin foliage like some enchanted object. This time I turned right, setting an easy pace as I followed a succession of steepening zig-zags up the empty mountain road. "It's not the goal, it's the walk", said a Zen master who knew how to travel. I kept to a no-hurry step and before long, stride followed stride in tandem with my breathing rate.

Walking 'within the breath' walking becomes a kind of meditation in action. If you're too goal-orientated on your walk, your mind's in one place (at the top, or wherever) while you're body's in another (here). You begin to walk 'outside' your breath…pain comes. Of course, it doesn't always happen as perfectly as I'm telling it: you might be with someone who's forever naming plants or birdcalls. You yourself might have a headful of mental chatter – the list of potential distractions is endless.

Dharmananda had described how to get to the temple and I halted by a bend when I spotted a notice in Hindi pinned to a tree. Just before it was a stone bridge spanning a dry streambed with an indeterminate trail wandering away through the scanty undergrowth. Then I remembered the swami had emphasised "many" notices and a well-marked path. Ten minutes later I reached several trees bearing hand-painted metal signs in Hindi; an obvious passage through the bushes and polished stones underfoot pointed the way. Overlooking the trail, a group of monkeys began to chatter amongst themselves as I approached; a large male with red eyes stared speculatively at my bag as if calculating its contents. I was so watchful of these audacious thieves that I almost didn't see the motionless human figure eyeing me.

The grey-bearded beggar was sitting cross-legged with his back to a sprouting stump at the side of the footpath. Dulled eyes stared before him as, in response to my one-word query, he raised a bony arm for a moment to indicate the curtain of monsoon forest which draped the 5000-foot mountainside above us. I nodded my thanks and dropped a coin into his empty metal cup. The monsoon was in fact several months away and the undergrowth, though plentiful, was dry and had none of the dank, claustrophobic inhospitability of, say, a Brazilian rain forest. With the early morning sunlight filtering gently through its motionless leaves, it conveyed no hint of hidden dangers. 'Open jungle', it was a welcome break from the ashram.

I cannot pretend a botanist's knowledge of forest plant life, but I was struck by the colour and form of some of the flowers. One small variety in particular, had heads so tiny they might have belonged to a fairy garden, while their colours were hauntingly subtle tints of pink, violet and yellows. Brownish cliffs, their faces too steep and ledgeless for trees, occasionally jutted through the vegetation like strange heads from long ago. Once, I passed before a canopied overhang with a delightful spring in its vicinity. A shallow fire-pit containing a few blackened twigs signalled some past occupancy. Further on, the trail steepened up a crumbling, green-hung ravine, where a number of well-scoured channels slanting down from a higher section of the path, suggested feasible short cuts for anyone with a head for 150-foot scrambles. For a moment I was tempted to introduce a little airy climbing into the walk, but I reminded myself there was no hurry and the feeling subsided as quickly as it arose: there would be plenty of opportunity for such ventures when I headed on from Rishikesh.

Despite my recent short bout of sickness, I found no problem in keeping to a steady pace, generally nudging the outer edge of the path so as to take full advantage of the occasional sweeping downward view when a gap opened in the foliage. One man, an unladen porter, gave me a glance of curiosity as he trotted down past me, heading for the road I had just left. Otherwise I saw no-one for the first hour. My boots, soon covered by a film of grey dust were a snug fit and scarcely heavier than the light plimsolls which porters tend to wear. Some Westerners trek in sandals, but this footwear has its limitations, not least the lack of ankle support. Also, they are often made of 'comfort' material which holds water like a sponge. Then it's like walking with a piece of wet fish tied to each foot.

The regular rhythm of steady walking soon produced positive results: a slowing of the onward flow of thoughts, an inner-outer balance and, finally, a feeling of great calm in which my senses were extra-finely tuned to their surroundings. Before many minutes had passed I rounded a corner and found myself approaching another natural cave in the limestone cliffs. This one was occupied by a seated figure. My pace slowed, remembering Dharmananda had referred to a forest cave-dweller who had not uttered a word for 11 years. Could this be he?

I drew closer, attracted by the man's rock-like stillness and the peaceful energy which radiated from him like a benign electric current. His shoulder-length hair and beard were as white as the Himalayan snows, but such was the vitality shining from his eyes, that it was anybody's guess as to his age. This was his chosen hermitage it seemed, and the woven blanket of subdued stripes thrown across his upper body served only to emphasise his reclusive calling. The cave's rear wall sloped backwards and was partly veiled in shadow, but I glimpsed a cloth bundle that might have been a rolled up bed and a few simple belongings tidily arranged. Not wishing to intrude, I paused before the neat arc of stones a few yards in front of him.

I could detect no rise and fall of his chest and such was the steadiness of his look that I thought he must be in some kind of trance. However, though his eyes did not move or show the slightest inclination to blink, I had the feeling he was fully aware of my presence. I felt disinclined to interrupt the cave-dweller's solitude, but the thought struck me that had he really intended to disengage himself entirely from human company, he would hardly have chosen such a semi-public spot in which to practise his spiritual exercises. Others apart from myself would surely pass by during the course of the day.

My hesitation must have communicated itself to the hermit, for I saw his eyes make a tiny sideways shift, to rest momentarily upon a flat-topped stone a yard or two to my left. That was the totality of his outward movement, but his eyes radiated a welcoming smile as I slipped off my shoulder bag and lowered myself on to the cool surface of the seat he had indicated. By the time I had arranged myself in a comfortable position, his head had resumed its forward direction, while the amused light in his eyes had softened to a deeper intensity. There was no way for me to know how long the cave had been occupied by this ascetic soul, but two things were abundantly clear. Conversation was not on the agenda. Secondly, I had no doubt I might learn something of value by dallying a while in his company in which I was reminded of the Chinese saying: 'Those who speak do not know; those who know do not speak'.

For several minutes I watched his motionless face: unlike my own, his eyelids never once blinked – or looked likely to. This fact impressed upon his countenance an air of absolute understanding and insight into whatever profound matters he chose to consider. Some subtle transmission seemed to flow from him to me and before long my own cherished views, opinions and encrustations of habitual thought seemed to drop from me like shards of broken pottery. It was as though I had opened to another, truer dimension of my being. A tear rolled from an eye corner and I felt that at any moment I would shatter the cave's peaceful solitude with an almighty peal of laughter. As I struggled to contain myself, however, the yogi continued to sit in silent and unpretentious dignity, his hands folded within his robes and his mind plunged in a profound stillness, utterly impervious to the idle antics of a passing stranger. Fortunately the urge to laughter soon left me and I quickly recovered my composure.

At the hermit's back I spotted a motionless lizard, its strange limbs spread-eagled across the uneven wall. Even as I watched, it appeared to grin, blinked, and began to descend the wall, first by an incipient crack, then more boldly as it neared the ground. The yogi moved not a muscle or flickered an eyelid as it scuttled over the very hem of his blanket, pattered across the book which lay open before him, until finally it slithered from view at the other side of his upright torso.

Stories abound in this part of the world of yogi forest-dwellers who suffered no harm within their circle of quiescence from the wild animals which roam there. In *A Search in Secret India*, Dr Paul Brunton tells of one such tale:

Once, a renowned monk called Ramana was sitting outside the entrance to his cave one day when a large cobra suddenly emerged from the rocks and stopped before him. When it raised its body and extended its hood the monk remained motionless. For some time the two creatures – man and snake – faced each other, eye to eye. Neither moved. Eventually the snake retreated without harming him, even though it was within striking distance. The cobra seemed to know he was harmless and left him alone.

Whilst this particular encounter happened many years ago, that is not to say these things no longer happen. I have heard similar stories from Western Bhuddist *bhikkus* when doing solitary retreats in the vast forests of Thailand. Wild animals, a Tibetan monk explained to me in Kathmandhu, were sometimes drawn to such recluses; somehow they seemed to sense their purity of purpose.

This doesn't seem too anthropomorphic an explanation to me. Back in the 'developed' world, I have sometimes wondered why a domestic cat, slowly picking its way around a group of seated people, pausing here and there before moving on, may eventually settle in the lap of one person, while ignoring all the rest. It has sensed something – but what? It is unlikely that the person's clothing smells of fish.

To return to my encounter with the cave dweller. The experience was not one which I could commit in its entirety to memory or jot down on a scrap of paper. Least of all, except indirectly, could I communicate it intelligibly to another, with my over-used rational faculties. Whatever the yogi had to give, it was not my mind alone which received it.

Eventually, the sound of approaching voices pulled me back to the unyielding world of plain truth, not to mention the rising heat. If I wished to reach the temple before the sun was at its height, then it was time for me to head on.

Wishing to leave some small token of appreciation, I fished an apple from the bottom of my bag and placed it silently on the rock I had just vacated. My action did not go unnoticed. The yogi's head did not move, yet from his eyes, a ray or *flash* seemed to dart from him to me. It was a thank you and a warm goodbye. Now that the approaching voices were almost upon us, no useful purpose would be served by lingering further.

The trail continued steadily upwards, a regular stony zig-zag which sometimes, when I glanced higher, would reveal its course by a faint furrow through the trees or a slanting indentation across a vegetated cliff-face. Once I heard a single loud cry or shriek from a tangled ravine on my left. I paused in my tracks for a moment or two to listen. However, the cry was not repeated and the silence settled down as before.

Overtaking a struggling group of Indian tourist-pilgrims a man addressed me enthusiastically:

"Birmingham, Notting Hill, Dublin – you are from London? Iron Bottom very fine cricketer."

I shook my head, hoping to avoid further interrogation. He was undeterred and, indicating a splattering of cow shit which had obviously been committed within the hour, he told me I should be very careful. His white-draped arm swung in a warning gesture: "Elephant, sir…many wild animals – we have been told!"

I gave the trailing, olive-coloured puddle a perfunctory glance. "Maybe."

Time passed.

Another gap opened in the greenery revealing a bleached-out, sun-seared world far below. Rishikesh looked filmy, vague, impalpable, too small to be a town; in the burning light, though, the Ganges wore a sheeny look of certitude, of things to come and events past. Dust rose from under my boots as I trudged on.

Before long the presence of a few broken tree stumps amongst the undergrowth announced I was approaching human habitation once more and the highest point on my climb. Presently, I emerged from the forest's cool shade into the full glare of the sun, whose light bounced harshly from the scrubby rocks, hurting my eyes. Several grey-skinned lizards darting across a glittering shale bank caught my attention: they had an air of pre-historic menace, and I was glad none were more than a foot or so in length.

I was now at my first goal, the broad ridge I had spotted some three hours earlier.

Below to my right was a neat patchwork of small fields with two women bent over, tending the ripening crops. I stopped at a fork in the trail where I drank the last of my water. Just then, I was surprised to hear approaching voices. They were raised, it seemed, in heated discussion, with a male American drawl being most conspicuous, with its trying-to-be-tolerant note. Shortly, two men and a young woman appeared, toiling up an airless, U-shaped rain-gully towards me. All three looked hot and vexed.

I recognised the first two immediately as the Californians who had recently moved into the ashram.

"Going somewhere?" I observed, as they struggled nearer, breathing heavily and arguing with their companion, a perspiring sadhu of pensionable age. My throat was dry despite the drink and my voice sounded like a bag of walnut shells.

Beneath a cotton hat against the heat, the girl's face lit up with recognition. Chestnut hair tumbled from under the headwear, whilst a pair of loose-fitting tie-up cotton trousers swirling with tropical sunset colours also caught my attention. Her tall companion wore a headband over near-shoulder length Nordic-coloured hair; his cheeks and chin were partly protected from the sun by a peppery, salt-rimed beard. A long-tailed embroidered white shirt over Nepalese trousers was somehow at odds with his pale, blue-eyed Viking-explorer looks which started me thinking of longboats pulling up on England's east coast.

His voice was taut with veiled exasperation as he huffed out of the gully to join me on level ground:

"Supposed to have been a peaceful night at the temple," he rasped, wagging his chin by way of a greeting.

I nodded back: "Something up?"

I sensed I was about to be drawn into a situation not of my own choosing. The sadhu swiftly caught up.

Sometimes travelling couples get too used to one another's company and a lone wanderer on the same road can seem like the way out.

The American's eyes narrowed into slits against the sun:

"We walked up last night; we wanted to stay at the temple an' see the sunrise...the guy came with us." He jerked the words out.

"-Snored the whole night!" his girlfriend panted in disgust, throwing a glance at the bearded holy man, as if he was some ruffian in an advanced state of leprosy.

So that was it. I nodded again as if I'd known all along. I was at the edge of the path, but there was no escape that way unless I fancied a fifteen-foot leap into thorn bushes.

He'd told me his name at the ashram – Henk or something- they'd been travelling for over six months.

"Where they put you up has no windows," she continued, clearly still aggrieved.

A second interruption: "The tiniest door, Jesus! An' when everyone's in they shut the door because of animals. It's a ten by ten… low ceiling an' after two hours we both had headaches we could hardly breathe."

He ran his tongue along the fringe of his moustache:

"About three o'clock the guy crawled outside. We threw his things after him and shut the door. He hollered for a spell then went quiet. That way we got some sleep." He glanced at his girlfriend for confirmation.

"The guy really stunk – Yuk!!"

I'd come onto the mountain for some peace and was already beginning to find the couple's tirade against the snoring sadhu more than a little surreal. Though short, he was a man of sinewy strength and carried himself with no little dignity. But were it not for his obvious sense of some great personal injustice, the hoary visage and vermilion turban canted at an angle over his brow, would have struck me as comical. I thought I remembered him elsewhere, perhaps in town in the company of other Westerners. He ignored me completely at first, his attention being fixed solely upon the target of his ire, the American couple.

Dust rose from under his feet as he danced up and down on the trail in a rage:

"This is not television!" he yelled, brandishing a rusty trident in their faces. "This is real! If there is something wrong – you should say so!"

"The guy followed us all the way," Henk confided, his head towards me as he coolly ignored the protest.

The sadhu seemed to notice me for the first time:

"I was sleeping at the temple and had to go out." His eyes blazed. "I return…door is closed against me - bag outside also."

He gestured at the mantle of forest to the north:

"Many wild animals in Himalayas. I sleep outside all night!" His singed moustache twitched violently as he shot the other two a look which said it all:

"You should speak me first – you not at the cinema!"

Henk affected a weary look and raised a warning finger to his lips as though placating a small child:

"Shu-ush, *baba*!"

"The *baba's* a bit crazy," his woman informed me flatly, as if the man didn't exist.

"Too much *charas*." They both began to laugh, an open invitation for me to take sides.

At the ashram I had formed a favourable impression of the Californians, but now I only felt uncomfortable in their presence. I shifted my attention to a glint of roof-tops four or five hundred feet below us. Casually, I changed the subject:

96

"I guess that's where I'm heading…the temple. I'd better get going…this heat."

"Yeah, it sure is hot!" each echoed the other.

At that, I abandoned the three of them to their own devices and started down the loose runnel up which they had just laboured.

Hardly had I reached its foot, however, when a shout caused me to stop and look back.

From the rim of the gully the sadhu, a bizarre blend of Eastern magician and rascally pirate, waved and began to descend towards me. The others were nowhere in sight.

With some misgivings I waited for him to catch up.

The holy man was only an obstruction if I allowed him to be. As Sabramuniya the spiritual guide said, it's all about timing:

"If you have the feeling that you're somebody going somewhere, that you're the one that's moving and everything else is standing still, then your timing is poor."

I took a strategic upwind stance as he approached.

Anger, which had earlier contorted his face, was now replaced by a kindlier look. Even so, I sensed he might want something off me. However, first impressions had been wrong in the past, so I decided to give him the benefit of the doubt. He obviously wanted to talk.

"Americans!" he shook his head, revealing knowledge of international geography which I would never have suspected, "they always think they are on Broadway!"

He proceeded to tell me his side of the conflict:

"Those very ignorant people…In our country when you stay at someone's house or in temple, there is proper behaviour. If there is woman and you get thrown out, it is very bad. Now everyone will think I am touching that American woman and it is not true!"

The skin of his lined face flickered with indignation.

"I am holy man. For us these things are big dirt, disgrace!"

I listened patiently without comment, though meanwhile the sun was making itself felt on my neck, and my forearms had long since ceased to sweat.

Suddenly the holy man switched topics:

"You staying Nikatan?" he asked confidentially.

I nodded, feeling the salt lining the edge of my lips.

"Don't like that place. Much money…doors close early; very strict."

"No smoke?" I added with a half-smile, noting the telltale signs.

A roguish chuckle rose from the holy man's stringy throat; his free hand dipped into his Western-type cotton shoulder bag.

"You want *chillom?* – I have!"

There is no confusion like *chillom* confusion and I shook my head.

A look of disappointment crossed his face.

"Tomorrow we visit other temples – I show you."

I didn't reply.

"I know of very special place…famous *baba* there. He only talk you if you go with me. We take taxi…only one hour and we are there. See famous *baba*. He perform miracles."

Though I was not uninterested in meeting some of India's premier holy men, I shook my head at this offer of a lifetime:

"Tomorrow I go Rishikesh…visit bank… I no need guide."

"It is no problem," the sadhu wagged his head, obviously much practised in the art of getting his own way. "We both go bank. Then take taxi…see *baba*. I wait you Swarg Ashram."

His eyes twinkled merrily. And, without waiting for my response, he turned and began to climb back up the gully, his bundled dreadlocks bobbing with every step like a bag of ferrets. I could see he was going to be a difficult customer to shake off.

Some of India's wandering ascetics, in possession of a finely-tuned sixth sense for spotting Western innocents abroad are often only too eager to make their acquaintance. Posing as a guide or spiritual mentor, the charlatan looks much like the true roaming sadhu at first, so the odds are invariably loaded in his favour.

I started down once more, the trail contouring the hillside for a short distance before dropping more steeply through low bushes to a clear stream bubbling down the centre of a fan-shaped valley. The running water was music to my ears and I paused for a drink. To the south and east rose a succession of blue hills cloaked with trees; narrow valleys shimmering with a silvery haze lent the scene a magical charm which was in no way diminished by my dusty state. The Himalayan glaciers and the eternal snows of the world's highest peaks were somewhere to the north, hidden from me as yet by the barrier of forested ridges which rose one upon another like a succession of giant waves. I promised myself I would see them before many more days were through.

From this point the trail carried me swiftly down the hillside of loose shale and I was soon at the edge of the tiny village I had spotted from above. All at once the mountain world of flinty footpaths and uncertainty was behind me as I proceeded along a cool lane with a succession of gaudy trinket shops lining its right-hand side. I don't know if it's a proven truism that the higher the village the more peaceful it is, but Neelkanth, perched at the confluence of two mountain streams, struck me immediately as the most tranquil hamlet I had so far encountered in

India. Of the few strollers who were about, none were Westerners, which was just as well, since the intrusion of familiar accents sometimes strikes a jarring note when one is immersed in the ways of a far-flung foreign land. Too soon, one is reminded of imminent return to the damp climate from whence one came.

The cheap and cheerful stalls soon gave way to a clutch of deserted restaurants, one of which had a mature tree growing from its stone-flagged floor and out through the tin roof – a clear instance of man and nature in peaceful co-existence.

I settled for a wooden table whose position gave me an uninterrupted view of the distinctly peaceful lane, a sublime panorama of receding blue hills- and, more to the point, a clear sight of the menu board. It seemed like happiness to be in that place, a blessed spot overlooked by the temple I had come to visit, its whitewashed walls a shining beacon in the flawless sky. An Indian with oiled locks and clad in a flowing *kurta* appeared for a moment from a narrow opening in the gleaming whiteness.

The return trek was straightforward enough, though after the lethargy of a two-hour stop, regaining the ridge proved a tiring exercise. There was no hurry though, and I took my time. Once, struck suddenly by the fantasy that Greystoke might be holed-up nearby, I halted and gave a loud jungle call through cupped hands. Apart from a manic screech some way off, the Green Hell did not respond. I don't think it was him, somehow. Anyway, Tarzan was in Africa. Jesus – the missing fifteen years –? Jesus…Tarzan? It's this heat, God –

I had only eaten a couple of samosas at Neelkanth and began to feel hungry again as I neared the cave-hermitage. Frankly, my thoughts were more concerned with bodily comforts and, unlike before, I didn't feel inclined to linger. However, I need not have concerned myself for the cave was empty. The arc of stones was still in place, but of the yogi, I could see no sign. He had gone; and so had the apple. I stared into the cave…felt an inner trembling on the edge of a breakthrough. Then I was descending again.

Chapter 6

I Meet a Man With an Ambition in Life

I had hoped to eat my evening meal undisturbed in a far corner of Choti *Wallah* restaurant, but before long I spotted a mournful face winding its way towards me between the tables. Resigned to the fact that I was going to be *in cognito* no longer, I forced a patient smile as the Dutchman raised his hand. His feverish gaze flickered over my face, as if desperately searching for something:

"I follow you into dis place – its okay if I join you?"

I gestured non-committally to the other seat, stabbing at some loose peas as I did so. Spiritual seeking at its most tiresome now confronted me.

The Dutchman had turned up at the ashram a few days earlier and had introduced himself to me in the dining room queue that same day. Around twenty-eight or nine, his neutral brown hair already well receded, was as damp as his perspiring brow. Hot eyes grilled me from a blotchy face which looked as if it had been attacked by calamine lotion; earnestness radiated from him, threatening to overwhelm all who came near. He seemed to have no awareness of the teeming world beyond the immediate vicinity of his own consuming needs.

By the end of the meal I was drained of all energy.

"I been at Uttarkashi," he informed me gloomily, his breath fanning my face like a picnic burner. "Yah, dos *babas* none of them knew anything; a complete waste of time!"

"Oh yeah?" I carried on chewing.

"In der south, Sai Baba – I went to see him. Two weeks I was der an' all I got was der shits…some people were very sick. Two weeks I was in der front row – the guy never cured anyone. Sai Baba couldn't cure my problems. He's a fake! All I got was der –"

"Listen, if you don't mind," I shot back. "Gurus don't exist just to make <u>you</u> feel good, to gratify <u>you</u>."

He wasn't listening.

"All those adoring Westerners… I should have stayed at home."

I struggled to remain civil. "Look – he might be all right for them…the other people. He can help them to face –"

The Dutchman interrupted again: "Who <u>are</u> you? I don't know who <u>I</u> am!"

There was a trace of despair in his voice, as he continued:

"I need to find someone - a master who can open up my mind...open my heart; solve all my problems. I been looking all der time for one. I have to find someone who can tell me what to believe!"

Suddenly I felt a pair of eyes turned our way.

Drinking a *Lassi* at an adjacent table was another ashramite, a self-contained Brazilian in his early forties. We all knew each other by sight.

He stared sharply at the Dutchman:

"Are you a little crazy? You think Sai Baba is there just to cure you of Diarrhea!? Look, my friend – what's coming across from you is that you have a very busy mind...Tell me, what makes you think that your suffering is any greater than anyone else's in here – tell me!?"

His words went unheeded.

"I can't talk to anyone in der Nederlands...I go all der day without talking. Only in India can I talk."

The Brazilian's voice softened: "Okay, but listen! You don't need a guru, a special person." He glanced about: "That woman, she is your guru...that little child can be your guru. This guy. Always you are searching outside of yourself: another person, books, ashrams – you don't need them. You have to take responsibility for who you are...Seek inside. Become your own guru!"

I thought the South American was talking sense.

The Dutchman shook his head. "What you tell me doesn't mean anything – it's superficial!"

The Brazilian replied with controlled passion.

"I will tell you one thing. Always you want to argue, ask questions... QUESTIONS – you have got to get out of your head: Understand!?"

He glanced at me, confiding, "He's not tuned in to what's going on."

The Dutchman's face glistened in the heat as he grabbed the bottle of *Limca* which a waiter placed before him. His eyes transfixed the Brazilian:

"You can be my guru!"

"I am NOT your guru! You say you are searching for someone- but you never listen!"

"Meditation," I ventured, "you should try it. Slow down."

"I done meditation for ten years... I tried all kinds... and what has happened – nothing!"

"You haven't even started yet," the Brazilian commented quietly, "you haven't even begun to meditate."

The Dutchman gulped at his soft drink.

"Just sit still...there's nothing to do. Watch the breath come and go. Concentrate. That's all."

A surge of conflicting emotions tugged this way and that in his face, as he focused momentarily on what I had just said.

I thought I had got through to him. I hadn't.

"Meditation's dangerous…jus' breathing, it don't solve anything." A strange light rose in his eye as he pointed his short forefinger across the table:

"You can be my guru – I seen you walking around the place… I seen you throw bread to der fish."

I couldn't recall the episode and remained silent.

The South American spoke again:

"He isn't your guru. You don't need beliefs, a philosophy. What you need is faith…in yourself. You are OKAY. But right now you have a very low opinion of yourself, yes? Tell me. How do you see yourself?"

For once the Dutchman didn't reply.

"Are you a mouse…a little shit? If that's the way you see yourself, you will be. If you see yourself as divine – that's great because you are! We all are. When you see yourself as divine it's half the battle…when you stop asking all the questions – THEN you've started!"

The whirl of the overhead fan grew louder and more insistent.

"The answer is blowing in the wind," I said aloud to no-one in particular.

The Dutchman fixed me with a creased brow – "This guy, he's – what ANSWER?"

The South American gave a tiny shake of his head and picked up his drink with a sigh.

My gaze returned to my plate…it was time to be somewhere else.

Later I asked the Brazilian 'the unnecessary question'. "Chickens," he replied. "I export chickens."

That night I had a strange dream.

Three of us were driving across India from east to west in a small, box-like car some years old. Sharing the front with me was a male painter friend from England. In the rear seat was the shadowy presence of a beautiful woman, though how she was related to either of us was not clear. We travelled effortlessly for some distance, passing through a number of small desert townships on the way. The air was very clear and everywhere I looked, physical objects shimmered with a radiant light. I saw a pile of rocks, orange, dark brown and purplish rising from the waterless sand – an ancient village? I couldn't be sure. My hands gripping the wheel, heartbeat, also my breath and the creaking of the car's suspension – all were of the same flow of energy. The rocks glowed in the landscape under the descending sun. To die seemed impossible. Towards evening we came to a smaller place, a sprinkling of dwellings standing amongst trees and shrubbery. One or two foreigners were present as well as local people, but their reason for being there was different. The scene switched. I was alone in a spacious garden filled with a variety of distinguished guests. A convivial atmosphere prevailed. Party-goers swarmed everywhere, but my task was to ask a question of 'a tall Indian dressed in pale green'. Where to find him, I had no idea. Two Western travellers hurried past with blank faces. As I crossed the lawn they vanished over a low slope. At the fringe of the party I came to a dark pool containing reflections of foliage and patches of sky. At the far side a heron stood motionless on one leg. Suddenly I remembered again why I was there and produced a document – a script of some kind – from my back pocket. At that moment, the man in green appeared at the top of a short flight of steps on the far side of the pool. He was standing centrally, very visible to me, but none of the other guests seemed aware of him. I walked, or more accurately, seemed to dance towards him, drawn by his aura of calm and certainty. Only a single step divided us when, for a split second, I looked away. When I looked back he had vanished. At that point the dream quickly faded.

The Dutchman knocked several times on my door in the next few days, but I lay still on my bed without answering and waited until he went away. Then on my way to breakfast one morning, I saw that the twin-doors of his room were flung wide open. That usually signalled only one thing. I took a peep inside as I passed, just to be sure. The narrow bed was pushed hard against the wall and a candle stump stood on a shelf. He had gone.

Uneasy in my mind for the next day or two, I wondered, had I done enough? Perhaps he'd only come to my room for a light? Did I back off him because his predicament mirrored my own in some way? I don't know, I really don't.

Meeting him triggered some childhood memories. In my early teens I read many annuals and adventure stories – explorers grappling with pythons, sailing the seas, or searching for lost cities in the jungles of Bolivia. True or otherwise, one such tale was set in the Superstition Mountains of Arizona. It was about a fabled gold mine. The original discoverer of the deposit, an old prospector from Holland, died without revealing the mine's precise location. The legend told how all those who went looking for it never returned. Pretty soon of course the mine became known as *The Lost Dutchman*. An obscure connection, but there it is.

I started to think about reincarnation. Perhaps it is intended that we cannot remember the details of living in previous existences? Think of the potential chaos of remembering everything from one life to the next!

The idea of 'coming back', of returning to this world in another physical body, is a strange one for many Westerners. In an age when "seeing is believing", we often hear that reincarnation is only another "belief system", a sentimental invention of those too timid to face the "fact" that death is not a door to another world, but the grim finality for all. Yet even the great Einstein felt drawn to observe:

"I believe that the present fashion of applying the axioms of science to human life is not only entirely a mistake, but also has something reprehensible about it."

To the advocates of 'one life and one life only', the terrible injustices of life, such as a baby being born blind, must seem inexplicable and indicating the absence of a caring Creator. But though it is difficult for the believer to offer proof of reincarnation which will convince the out-and-out sceptic, there are numerous examples of Western thinkers who have been open to the possibility. In his novel, *Siddhartha*, a story about the search for spiritual truth, Hermann Hesse had reincarnation much in mind when he wrote,

"He saw all these forms and faces in a thousand relationships to each other… None of them died, they only changed, were always reborn, continually had a new face; only time stood between one face and another."

Carl Jung in *Man, Dreams and Reflections* said, " I could well imagine I might have lived in former centuries and there encountered questions I was not yet able to answer; that I had to be born again because I had not fulfilled the task that was given to me."

For 'popular' consumption, a variety of "irrefutable" proofs are widely available in Western bookshops. Their evidence is drawn chiefly from accounts of near-death experiences, out-of body descriptions and the cases of people who regress to earlier lives under hypnosis and detail the surroundings in which they once lived.

A.C. Bhaktivedanta Swami Prabhupada, however, warns of their superficiality. None of them, he says, answer any of the fundamental questions about reincarnation: Does one reincarnate time without end or only once? Is it possible to control our future reincarnations? Can we be reincarnated as a plant or an animal? What is the relationship between *Karma* and reincarnation? Little serious attention is given to such questions, he says. Where, then, are we supposed to look for scientific evidence?

Swami Prabhupada directs our attention to the Vedic knowledge of India, particularly the pages of the Bhagavad-Gita. What Western science is now finding, he tells us, the Vedas had already explained many centuries ago:

"The basic principle in the quest – that I am not this body – is the root of man's dilemma."

Unfortunately in a book of no more than moderate length, I can do no more than skim the surface of such a vast and compelling subject. I can only hope that the still-unconvinced reader will be sufficiently open-minded to peer through the microscope of some of the other great teachers whose names I have sprinkled all-too thinly amongst these pages. Shortage of space precludes the mention of many more who have made their profound contributions to the subject. Having said that, there are a number of Vedic cosmic laws which have puzzled me for a long time and to which I shall now briefly refer.

In physics it has long been known that action and reaction are equal and opposite; it has been observed that nature behaves that way, neither normal nor personal...just the way it is. There is no need to disown one's own cultural heritage to see that, logically, this law might equally apply in the realm of one's own behaviour too. One does not have to give up being a Christian, a Buddhist or stop reading the Koran to recognise the possibility.

Vivekananda says: We, we and no-one else, are responsible for what we suffer. We are the effects and we are the causes."

In Sanskrit, *Karma* means 'action', thus by expansion, a sequence of actions. It is the law by which every action, feeling and thought is finally returned to the individual who first set the wave in motion. Thoughts, no less than 'deeds' are actions according to *karma*. In daily life how we think and feel both creates a reaction in others and may even influence the physical environment. A useful guideline is that the *motive* which first set the impulse moving will decide the nature of the reaction. For example, a man who works hard to make money in order to provide for his family will generate a different karmic reaction from the one whose motive is to amass a fortune. The consequences are both immediate and remote. Every action of generosity, to give another example, slowly weakens the bondage to greed in our own minds.

Karma is not changed easily, but neither are our lives in the future totally determined by our actions today. It does not simply mean 'fate' or 'destiny' as many in the West appear to think.

We in the West are often guilty of self-indulgence concerning *karma*. On the road to Manali I shared a restaurant table in the quiet town of Kandi with a Western traveller in his early thirties. He had been on the move for three years and explained away his inability to settle in one place by reciting the mantra,

"It's my *karma*: I have to travel. Perhaps in the next life–"

It sounded more like imprisonment than freedom.

I found it a hopeful sign to discover that *karma* is by no means forever fixed and unalterable, the human being the poor victim lashed to the wheel of a heartless, inescapable fate. By our thoughts, feelings and actions today, the Hindu texts tell us, we are modifying our future at every moment. By means of self-cultivation, patterns of delusive thinking and behaviour are gradually replaced by more harmonious traits within ourselves.

In traditional Buddhism, on the other hand, these patterns are passed on through a succession of rebirths, though it is *karmic energy* which is transmitted rather than a reincarnated personality. There is a shift of emphasis. Instead of a permanent soul ('never born, never dies') following its lonely path through numberless rebirths, there is now a collection of properties or aggregates (the *skandhas*). These include: the aggregate of feeling; the aggregate of mental factors (fifty two kinds) and the aggregate of consciousness. All is changing at every moment and the function of the *skandhas* is to determine *karma*. When one life is finished *Karma* creates a new collection of *skandhas* e.g. another being.

This Buddhist teaching has always seemed a little obscure to me. If there is no permanent soul why does *karma* appear to be immortal? Is there some truth too deep for the intellect to follow? And who or what is it that at last enters the ultimate goal of Buddhism, *Nirvana*? Buddha is said to have maintained a "Noble Silence" when asked questions of this kind. Perhaps when we stop asking, always asking the answer comes… In silence. Or is there no answer because by then there is no question?

Many resent the concept that 'we, we alone', determine our fate, that destiny according to the karmic law is ultimately self-earned. They cling to the idea of a Great Provider to whom they can turn in time of need – or hold responsible when things go wrong. Fortune, Hinduism tells us, is a turning wheel; that power which brings us problems will also carry them away.

Who is right? Is the system which places the *Brahmins* at the top of the caste system and the Untouchables at the bottom, less commendable than the Christian or the Buddhist way? Or are all 'explanations' of a similar order: the futile mumblings of finite mind trying to make sense of BIG MIND?

Maybe we weren't supposed to know that for 500 years after Christ *Karma* was part of the Christian faith. But for the prejudices of a 4th century convention of bishops in Constantinople, a gathering of 'experts' who condemned it to exile, *Karma* could well have become a pillar of modern Christianity! Emperors condemned people to death in those days for believing in reincarnation. In 'enlightened' Europe centuries later, heretical books were ruthlessly destroyed and numerous churchmen were burnt at the stake for holding these heresies.

Nevertheless, we can still turn to passages in the Bible which seem to show that Christ and his followers were not unaware of reincarnation. In the story of the man born blind Jesus is asked by his disciples, "Who did sin, this man or his parents? If not the parents then did the man sin in the womb before birth?"

One would hardly think so. The reference must have been to a previous life.

Where do all these tortured questions leave us? Which religion gets the prize for being 'true' whilst all the others are 'misguided' at best? Popular in modern times is the syncretic path in which a personal combination, a cobweb of bits and pieces is made from the different religions. Some call this 'New Age' but it is little more than what the Gnostics stood for in the early days of Christianity. Gnosticism was ultimately a doomed concoction of teachings borrowed variously from Greek and Eastern sources along with the contents of the Bible. Though popular in the beginning, it had no living force as its centre. As a bundle of high-minded truths it only made sense to the initiated few and eventually died out.

By and large the Western visitors to Indian ashrams aren't seeking to be embalmed by the ancient lessons of an exotic religion; it is more a quest for a transforming methodology. The real choices are within oneself.

Will there some day be a great synthesis of the huge diversity of religious doctrines, each thread uniting with its neighbour to create the Greater tapestry?

That master of the everyday parallel, Swami Vivekananda, wrote:

"When you watch a kettle of water coming to the boil, if you watch the phenomenon, you will find first one bubble rising, then another and so on, until at last they all join, and a tremendous commotion takes place. This world is very similar. Each individual is like a bubble and the nations resemble many bubbles. Gradually these nations are joining, and I am sure the day will come when separation will vanish... A time must come when every man will be as intensely practical in the scientific world as in the spiritual, and then the harmony of Oneness will pervade the whole world. We are all struggling towards that one end through our jealousies and hatreds, through our love and co-operation."

Clearly, depending solely on one's own efforts for salvation can be, is, a searching test of maturity. Though the New Testament says, "Whatsoever a man soweth that shall he also reap", how can we know for what deeds in an earlier life we are being rewarded or punished? If 80 per cent of the people in India are poor, does that mean that 80 per cent have done wrong deeds in this life or an earlier one? And what of whole communities wiped out by floods during the annual monsoons? Is there group and national *karma* to account for those events?

I have heard a few answers: "Some people are poor because the resources don't meet the demand". The caste system "provides a sort of welfare arrangement at village level. If people are in need they can go to their caste for help. Caste provides a place and a role for everyone". And so on.

And the Dutchman – would he be any less lost for knowing the answers to these contradictory questions? The gurus of India couldn't help him. Would he fare any better on the analyst's couch in Holland?

Buddha said, "Be a light unto yourself." In other words, nobody is a guru – so what could the psychologists tell him of any value? Maybe he would feel good for a day or two after a session. Then he would have to return…and return again. Would his analysis <u>ever</u> stop? In the West the psychologists have taken over from the priests as the mind police. Why are there so many psychotherapists? How is it possible to prove which one has the monopoly of truth?

In treating the Dutchman what would be the analyst's aim – to 'free' him? I rather think not. A cure would be to make him 'normal', indistinguishable from the herd. But it wouldn't work because the cause would still be there. He would still be in his mind. How can you be objective, scientific, if you are part of the subject under observation? There is no distance. It is like studying fruitcakes. If you want to study fruitcakes, it is imperative that you should not be one. Fruitcakes studying fruitcakes is nonsensical.

On the mind, Osho, formerly Bhagwan Rajneesh, wrote:

"Western psychologists have not even pondered over the fact that the East has created so many enlightened people, and none of them has bothered about the analysis of the mind. Just as in the Western literature…there is no idea of going beyond the mind, in the same way there is nowhere in the Eastern philosophical literature any mention that psychoanalysis is of any importance."

Meditation leads us out of the mind; it takes us beyond questions and answers. Yet at present few psychologists are prepared to risk their reputations by acknowledging the potential benefits of meditation. You may, through its practice, come to know at a very deep level that nothing is permanent…All that arises passes away. It doesn't sound much of a discovery – 'all that arises passes away' – until you've dwelt on it for a while and observed how transient things really are.

And of previous lives – what of those?

I could not forget the Himalayan astrologer's interpretation of my own horoscope. He recalled accurate details from my early years before commenting:

"There are important aspects from your earlier existence which will dictate the kind of experiences you have in this life. In your past life you were thinking too much about material things. Yet it is possible that you were a monk in that life. That is why you are being drawn once more to sacred places now. The ashrams and monasteries, mountains and rivers – these are all places of power and joy."

As I listened another part of me heard Cassius' warning in Julius Caesar:

"The fault…is not in our stars

But in ourselves that we are underlings."

The astrologer's tone was matter of fact as he plausibly disclosed further aspects of my life and its needs.

"Nature is important to you. Also, looking out across large expanses of water, the ocean, is very good for you. Your mistake in your previous life was not to use your Venus power to spread what you know. Don't waste your time relearning what you already know from previous lives; you will only get bored."

It sounded simple.

One night, against the background of a thunderstorm, I had what seemed like a significant dream.

I was walking across a sunlit waste of sand and shining rocks towards an entrance in a high cliff. I entered a huge cavern whose walls gradually vanished in deepening shadows. I realised I was in some kind of library, for in every direction rose row upon row of impressively- titled books. A portentous quiet filled the cathedral-like space and I moved as silent as a cat's ghost across the sandy, stone-flagged floor. My vision was now adjusted to the dimness of the interior, and I saw the bent-over figure of a man working at a table littered with notepaper and precarious stacks of books. A hanging lamp softly illuminated the scene, its cord disappearing into the darkness overhead. The man was scribbling with great urgency in a large notebook and took not the slightest notice of me as I approached. I could see only part of his profile – but I felt sure that I knew him quite well. Slater? Williams? Perhaps he was a friend from the past with whom I had long lost touch? That mannerism of stroking the side of his cheek with a finger end...I wanted to peer over his shoulder to see what important things he was writing. There was a click; the light dimmed: something was happening to the books! Each and every one began to glow, feebly at first but with increasing strength, their spines gleaming with a succession of vivid saturated colours...emerald, citrine and glowing pinks and reds – the volumes in my immediate vicinity pulsed with a slow, hypnotic rhythm, as though they vied for my attention. The higher shelves appeared lifeless in comparison, though a few edges glimmered faintly in the pale light which fell from above.

Those shelves within an easy reach for a person of average height were labelled POPULAR KARMA. The condition of their bindings indicated they were much in demand. A wry smile came to my lips as I noted a title here and there:

A professor from an obscure Kansas university had written, *Karma and Your Earning Power in this Life*. Not only were India's own renowned mystics fully represented in the library, but it seemed that every nation on earth had its own galaxy of self-appointed sages only too willing to share a few crumbs of half-digested philosophical truisms with the world.

The dream shifted and I began to float up and up, past titles of a very different pedigree.

As well as books there were erudite papers and neatly typed treatises – all dealing with the one subject, *Karma*. Some of these were very abstract studies which I might have once have read with interest. Now I found them as dull as ditchwater and without life.

One or two, however, had a thoroughly practical approach and promised much solitary endeavour and virtuous behaviour if the goal was to be attained. I opened one of these at random and read:

"The *rishis* of ancient India analysed death as the withdrawal of the electricity of life from the bulb of human flesh–"

I was about to read further – when again the dream shifted. Books began to flow past me in an never-ending stream. A crowd's cheer rang out, fading away in a twinkling of lights. Suddenly I was back with the enigmatic figure poring over his books. The volumes began to vanish into a kind of fog. The figure glanced up as if aware of my presence, his gaze ran right through me and I realised I was invisible to him. Even so, I was struck dumb with instantaneous recognition. For the seated figure, weary of eye and surrounded by books – was myself!

The man – me – notices for the first time that a book has slipped from the table and stretches to pick it up. He opens the book and a look of surprise crosses his face. Except for a single black page the book is empty! Both sides are blank. A long pause follows while he absorbs this fact. Finally the man turns his head and stares as if into the void. A lighter emotion disentangles itself from within and brightens his face: I have the distinct impression that he has reached some important turning point. As this impression grips me, the glow starts to fade from the scene. The man's mouth opens as if he is about to speak, but no words pass his lips. A mist begins to occupy the cavernous reading room; shapes begin to lose their form. My energy is pulling me elsewhere and all at once I realise I am back in normal consciousness.

The dream left me feeling drained and after breakfast I smoked a *beedi*, a thing I never normally do. Why I bother, I don't know, since I never smoke at home. Invariably though, I have a couple of packets of 'the poor man's cigarette' tucked away somewhere when I'm visiting Asia. They don't even taste nice and their appearance is nothing to write home about: a thread of tobacco tightly wrapped inside a couple of dried leaves, and secured by a round or two of coloured thread. GURU BRAND is the one I usually go for. The seated figure's benign countenance on the conical wrapping paper seems to say, 'If I like them, you will too'. But I don't really. I'm inclined to think that some gurus aren't for everyone.

As the reader may already have noted, ashrams by their very nature attract a proportion of those individuals who outside in the 'real' world might draw the damning labels, 'misfit', 'solitary', or 'not management material'. It would be tedious to catalogue every encounter I had along the way, whether it be in ashram or boarding a bus in some far-flung scorching village. Nevertheless, there are those which stand out in the memory and are worth recording, if only to further illustrate a certain weirdness, which some foreign visitors possess in abundance.

One afternoon, as befits one who is neither the 'mad dog' of the aphorism nor has a mortgage to pay, I was resting beneath my fan, daydreaming about an ice-cream from *Chotti Wallah's*. Unable to sleep in the sluggish air, my eyes, half-closed, strayed to a dark, ambiguous blotch on the ceiling. Was it or was it not an insect which might at any moment drop uninvited on my bare legs?

As I was idly contemplating this crucial question, the light in my room suddenly dimmed as if curtains – there were none – had been drawn. At the same instant the window flyscreen creaked as if something heavy was pressing against it. Immediately alert to the possibility of an intruder, I forgot about luxury icecream and raised myself on to my elbows to look closer – I hadn't heard of anyone being robbed inside an ashram, but there was no harm in being vigilant.

I sat still, scarcely breathing at all.

Soft footfalls shuffled slowly away and I breathed more easily as the shadow vanished from my window. Once again I lay down. Two minutes later the same thing happened. When the creaking came, I saw a large pair of cupped hands ringed by a pallid face pressed up to the window. The face turned first one way, then the other. I raised a hand to brush an ant from my knee. At the same moment I heard a short exhalation of breath, the hands and face disappeared and the slow footsteps receded for the second time. I didn't care to be spied upon in such a blatant way; in an ashram I could see no reason for it. I stuffed my feet into my thongs and made for the door.

When I looked out there was nobody there. On either side were silent doors, all apparently shut.

Puzzled, I took a step forward. To the left only a pair of cotton trousers hanging from a line suggested a room was occupied. I stared out into the baking quadrangle. Nothing moved. Was it only my imagination? Perhaps I was about to go down with something serious? Suddenly the silence was broken by a voice as a very large man loomed from the next room but one to the right of mine.

Dressed from head to toe in a black monkish robe reminding me of a granddad's nightshirt, his head was several inches above the lintel of the door at his back. Around his thick neck hung a tight necklace of a devout's *mala* beads; his feet were encased in an enormous pair of old-fashioned sandals. His black hair was pressed flat across his temples as if he had just removed an ill-fitting hat or perhaps a turban. His skin was too light for him to be taken for an Indian, although his eyes were dark enough for a casual appraisal to mistake him for a member of that race. I stood my ground as he took three or four ponderous steps towards me.

People of such monumental bulk can be intimidating, but his voice betrayed him, carrying little of the conviction which his physical appearance projected.

"Hi," he drawled softly, "I'm yor new neighbour."

He reminded me of one of those part-theatrical Daumier figures posing with assumed authority on some imposing court-house steps. Despite the climate his face was unusually pale, like one who has stayed too long indoors or just that moment received bad news. He looked around forty, but when I looked more closely into his eyes I thought perhaps my estimate was a few years too many. His grip was clammy when we shook hands.

Now he raised one hand before him in the manner of a mystic master descended briefly from the eternal snows:

"I am very pleased to see you," he informed me in a curiously sluggish voice. "My name is Chaitanya." There was a moment's lull before he added: "but my real name's David."

My earlier caution subsided as a first glimmer of what this man was about began to dawn upon on me. Was he in possession of a spiritual master's unusual powers? I doubt it. I tried to recall the list of daunting saintly virtues which that revered name, Chaitanya, represented for millions of Hindus: to me it seemed preposterous – a presumption of the highest order – for a foreigner to adopt such a name as his own. If a guru had bestowed the name upon him, that was a different matter altogether.

Those who have found themselves in a similar situation will appreciate the modest dilemma which faced me. Just how seriously should I take this man?

I have to admit that his eerie and somewhat robotic physical presence quickened my decision to give him the benefit of the doubt for the time being. I didn't want to make any bold protests which I might instantly regret if the stranger was at all unstable. However, I should have acted quicker, thereby saving myself an extended period of tedium as captive audience.

Perhaps my body language signalled 'encouragement' for 'Chaitanya' pitched straight in.

Through scarcely moving lips he narrated his life's tale in a slow, persistent monotone which gave me no opportunity to interrupt. His delivery, fortunately, soon produced a trance-like effect on my own mind, and in this realm of mental torpor I fortuitously heard only a few details from his vast lumber-room of half-digested religious ramblings. He made no inquiry as to why I was in India or where I was from. Who I was did not concern him: he wished only to inform me about himself and his great ambition.

America was the country of his birth, his father being from that country, his mother being an Indian Hindu. India had been 'home' for many years now and he declared with some force that he would never again return to the States. He told me of several heartfelt goals he wished to realise. The first – and not such an unreasonable aim, one would think – was "to be accepted by Indians as one of them". Most of what he told me slipped swiftly from my head, forever beyond recall except under deep hypnosis; but one detail of this mysterious man's bizarre pronouncements stood out as an example of self-deluded grandeur. Inhaling deeply, he seemed to enlarge before my eyes until he resembled one of those top-heavy elder statesmen one sees on television:

"By the time I am 50 years old," he informed me, "I want to be prime minister of India."

He looked over my shoulder as though his Destiny was scorched deep across the vaulted sky for all mankind to wonder at.

"Good idea," I replied, my features creased with approval for his noble task. Perhaps I was setting my own goals too low?

'Chaitanya' continued gravely, adding a grand, ceremonious gesture or two which helped to sharpen my impression of a man rehearsing for a major role in Life's tragi-comedy.

A task he had set himself in order to further his ambition was to learn ten new words of Hindi each day. A yawn threatened to engulf me as he began to count from one to ten in that ancient language:

"Ék..Do..Tin...Namaste, méra nam Chaitanya hai." He reeled off a string of beginner's vocabulary to show he meant business. He had hired a 'real' teacher to help him in this scarcely plausible undertaking.

I shifted my weight from one leg to the other.

"Lemme see," he added, his mind struggling with the strenuous arithmetic, "what he tells me in the morning I repeat three times in the afternoon…Er, that's – by the end of the week, I know – erm, seventy new words."

He obviously had the qualification for high office.

"What might be your main policies when you eventually become prime minister?" I ventured, innocently enough. I was trying to humour him, not mock him. An unexpected feeling of compassion suddenly filled me as he threw out a hand in a placatory gesture:

"I'll fix things…New ed'cation programme – too much technical schoolin' over here… I'll tour the country…get rid of corruption…There's lotsa things I intend doin' when I…" His voice trailed off into a deflated silence. For the first time I sensed a wounded spirit.

Suddenly, as though he had just remembered, he swung his head in the direction of his room.

"Gotta get on with m'vocabulary programme now."

With a soft swish of costume drapery he was gone and I was left alone once more on the hard stone steps of the quadrangle. For a while I didn't move. Behind me, a sonorous chanting of simple Hindi had already begun. Slowly, sometimes hesitantly, like a slender reed pushed to and fro in the breeze, Chaitanya's voice swung back and forth, ebbed and flowed – checked and began again.

Somewhere – perhaps in the pauses – I heard a person in need.

It dawned on me that I could not simply dismiss such a type as a 'neurotic' in search of himself in a land far from home. After all it isn't so uncommon an experience to feel uncomfortable within the Western psyche. True, he appeared to lack all capacity for objective contemplation of his own thoughts and actions, but who was I to judge others so harshly? He was far from being the only one to have forgotten his true nature. His ambition, ludicrous though it might seem, was probably no more delinquent than my own, teenage fantasy of training to be a human cannonball.

I was quiet now. I gazed steadily at the flawless, near-white sky over the flat roofs. Its infinite space came closer and closer… I could feel myself being drawn into that space. The sky was beginning to move in all directions, enfolding me within it… I was part of the emptiness – was the emptiness of that white, white sky.

Later that afternoon I took a short stroll along the Ganga's edge.

It seemed clear to me that whether we know it or not, we are all following the same path. The expression of that Way might vary from one individual, from one culture to another, but the *Dharma* is the exclusive property of no one continent, culture or creed. How could it be otherwise? We may talk about the 'holy Ganga' but ultimately all rivers, all water is sacred. It is only our egos which draw distinctions between spiritual and non-spiritual.

So why go anywhere? Is it as Robert Louis Stevenson says, that the traveller travels, "not to go anywhere, but to go"?

There is that beautiful insight of Saint Exupery's in *Flight to Arras:*

"It is true that when we travel we are in search of distance. But distance is not to be found. It melts away…And escape has never led anywhere… In Pasteur, holding his breath over the microscope, there is a density of being… Cezanne, mute and motionless before his sketch, is an inestimable presence. He is never more alive than when silent, when feeling and pondering. At the moment his canvas becomes for him something wider than the seas… What we are worth when motionless, is the question".

I don't offer this as a theory. I think it is an error to assume philosophers always live according to their professed philosophies. Maybe they try to.

Many have suffered through trying too hard to live according to some great thinker's ideal.

Take one such man, Ouspensky. He was a profound thinker, a great man whose ideas have changed the lives of many people. But did he know himself? He was a poor driver, I read. He had numerous traffic accidents. Why so many? Did he admit to being a poor driver, a danger to others as well as himself? Quite the contrary, in fact. The accidents were caused, he said, not by his own mistakes: they happened because as soon as he climbed into his car he would be distracted by "psychic enemies".

In this case, are we supposed to believe him just because he was the great thinker? The story, even if it is fiction, reminds us that life can't be lived for long according to logic or some extreme ideal. Humanity flies out of the window when we try.

I halted within sight of the bridge. It looked like a rainbow, connecting two worlds. The river, glittering now under the lowering sun flowed on as it had done for millennia. After only a few short weeks, the Ganga had begun to take a hold upon me. Wherever I went, there it was. How can I express what I felt? I was beginning to feel it was part of me and I of it. As the river was forever changing, shifting its course, so I felt less and less attached to a narrow sense of identity. The immaculate river flowed on into the distance, emptying itself finally, into the ocean which receives all rivers, just as each individual must also finally dissolve in the universal when his time comes.

The swami was sitting in his customary lotus position on the threadbare carpet of the shrine room. There was nothing forbidding in his manner as he glanced about as, not for the first time, he recalled his introduction to ashram life:

"Before I decided to become a monk I was Indian army officer – very *pukkah* job! In this country such a position carries enormous prestige and respect. All menial jobs such as polishing shoes, cleaning latrines and so on, all are done for you. Most of the time life of Indian army officer is very easy; you can ride about in a jeep all day with your chums!"

He opened his arms in a mock-heroic gesture –

"And I gave all that up to become a simple monk!"

The laughter died away as he continued,

"But it was not so simple at first! When I arrived Ved Nikatan, what was the first job they gave me? Can any of you guess?…Toilets! They gave me a brush and made me clean out the toilets all day long. I was very proud man who had always left such jobs as cleaning shitty toilets to the cleaners. It was very long time before I could stop feeling resentment at being given such menial tasks… Our guru here, is very observant man; he knows exactly what kind of work a new monk needs!"

His voice softened as he smiled at the memory.

"So you see, it is only by tackling such jobs without resentment can we hope to diminish the strength of the ego and make spiritual progress."

His gaze fell first on one person, then moved to another. He raised a slender forefinger:

"Pain arises through our personal likes and dislikes. When we identify too closely with <u>my</u> feelings, <u>my</u> intellect, with <u>my</u> mind – then much pain and suffering will arise. When we say, 'I am not happy with myself' we are identifying too closely with the physical being, the body. We shouldn't allow our feelings to have such a free play with us. There will always be imperfections. Some things- people, certain everyday experiences – attract us. Others repel us. Likes and dislikes get established in our minds because we make judgements. But since we are limited human beings, judgements arising out of limited minds will inevitably create havoc with our personal and family lives. Opinions based on whims and fancies will cause powerful cross-currents in the mind and will bring tension and much unhappiness. Observe your own nature. People with strong likes and dislikes will never be happy."

He gestured to a latecomer, a stubble-cheeked Israeli, to sit down, before drawing on a humorous example:

"Say you are very thirsty, hot…The most important thing you want in the world at that moment is – an icecream!" He chuckled at one or two affirmative nods around the room.

"You have one icecream – very nice! You decide to have another, the first was so delicious. The second icecream is also very fine, but not so nice as the first. Your immediate desire for icecream is satisfied."

To general levity, he added, "You are no longer thirsty – taste buds are used to the novelty. A friend offers you a third icecream – you throw up your hands in disgust. You are now heartily sick of icecream! So we see how satisfaction of wants is quickly followed by pain. In the rest of life it is the same: in sex, when we start a new job, eat, buy new clothes, go on holiday – desires never get fulfilled. Desire One is followed by Desire Two by Desire Three and so on. Here in this ashram," Dharmananda's gaze roamed mischievously round the hall, "I see how quickly Western men make new acquaintance with Western women! Suddenly it is all very exciting. This man and woman – they have only eyes for each other. The whole world can go to ruin, but so long as they are together, everything is all right!"

A note of gravity crept into his voice.

"<u>But</u> – what is the very first thing that happens next, huh? These two think they are very much in love with each other."

The swami turns his glistening black eyes upon me for a moment. He strikes me as being supremely confident in his own knowledge, yet I can detect no trace of arrogance and am impressed by his patience and ability to command attention. I notice the gheko is back, watching the proceedings intently from high on a pillar, its gaze steady and unblinking. Somebody breaks the silence with a short observation.

Dharmananda folds his hands in prayer and directs a gaze of mock-despair at the high, shadow-dimmed ceiling.

"The first thing that happens is: *asanas, pranayama,* meditation, all those spiritual exercises discovered by the *rishis* of long ago – all fly out the window! The great science of Yoga is but a heap of ashes!"

I sensed a collective sigh of recognition.

"Where is this happy couple at six o'clock in the morning when yoga class starts? They are nowhere to be seen! Ten o'clock I go to their room."

He knocks on the wall of an imaginary stronghold.

"No answer. Door locked, shutters all closed. They are both inside snoring to glory... Yoga is very far from their minds."

I join in the burst of wild laughter.

"The-en," the swami pauses and a look of genuine sadness fills his eyes, "three weeks later, what happens?"

His voice slows and I know he has observed these situations all too often.

"Usually it is the man, though not always – after only three weeks, the man starts to get a little tired of this woman. He finds fault with her... His eyes, they stray to other woman who has just moved in next door. First girlfriend now very jealous...suddenly these two are quarrelling very loud – I have seen this happen so many times in this ashram!"

Dharmananda shakes his head and leaves the conclusion of this story to our imaginations. After a while he again begins to speak.

"So you see how this couple – who were so much in love – have not found peace – for the simple reason that their mutual attraction was based on body consciousness, their likes and dislikes. Only when you begin to get established in inner Yoga will you begin to move with the higher self and rise above personal preferences. That is where the guru-disciple relationship is so important. The guru cures the student from his rigid ideas of likes and dislikes. Eventually he sees that satisfying the senses is like being a puppet, a slave to *Maya.*"

The swami then turned his attention to the important features attending a sincere spiritual life. How might he have got on with Pope Paul IV who in 1964, announced to the world, "Honesty compels us to declare openly our conviction that there is but one true religion, the religion of Christianity"?

"Childhood; Adolescence; Maturity – these are the main stages in the spiritual life. Most aspirants fall into the second category. The *samskaras* from our previous lives rise like air bubbles deep from the subconscious and compel us to go higher. If we try to return to material enjoyments, we will then find we do not enjoy them as before. It is a process. So we must expect to feel conflicting pulls, both up and down. If the idealistic things are not happening in the street, it doesn't matter. There is a clash, but we must learn to stick to the path."

Aware that some minds harboured a sceptical question or two, the swami lightly tossed out a challenging aside of his own:

"Naturally, I do not expect you to agree with everything I say. Some Westerners who come to this ashram are out-and-out sceptics who want prior assurances that spiritual disciplines are not just a form of indulgent self-hypnosis. Such assurances are impossible to give in a way which will satisfy these people. You must judge these matters by your own experience. With persistence, you will come to understand that mystical truths are not just beautiful subjective feelings – they convey knowledge of the spiritual world. You know from your own studies that modern science is at long last coming to see that matter is not the primary reality."

He sighed and looked steadfastly amongst us. "I am not being critical of any of you who are here today, but there are those who use this ashram only as cheap hotel. They arrive from beach party at Goa. They come to one of my talks, hear something they disagree with – and leave same day, saying, 'That swami is rubbish!'"

In front of me the fidgeting Israeli muttered darkly under his breath. The monk's eyes turned towards the newcomer:

"If you are at all serious about the spiritual path, then my advice is – stay away from Goa! I understand why it is very attractive to Westerners…Maybe there are spiritual teachers there as well. But there are too many distractions. If you go to that place you will hardly advance one step along the way!" His finger pointed to the open door as if Goa, the place of dawn to dusk hedonism was no more than a twenty-minute bus ride away, instead of a gruelling thirty hours by train.

The swami's eyes softened. "Come to places like Rishikesh." He looked about him. "Can anyone tell me what this town is famous for?"

"The Beatles," murmured a man sitting behind me.

There was no hint of reproach in the swami's response:

"Ha! – I ask a serious question and that Englishman brings me down to earth with a funny remark! Yes, of course, until the Beatles came along in the 60s the West had not heard of Rishikesh. But to Hindus, this town has been very, very holy place for centuries. Here, people come with the specific intention of listening to spiritual master, of dedicating their time to higher qualities…No, in India, Rishikesh is not famous for dreadlocks, chillom smoking and all the rest. People will travel many days to get here because of the spiritual vibration which has built up over the generations."

A shadow of a smile crossed his face:

"Unfortunately, as well as genuine masters, there are always many fakes. It is often difficult for you Westerners to know the difference. Over the last twenty years or so, many of Rishikesh's real yogis have begun to find the place too busy, too commercial. They have moved away, higher into the Himalayas, where they will not be disturbed. It is problem for people in the West to understand just how much influence these yogis can have upon the world – when from the outside they appear to be doing nothing. Yet if suffering is to be reduced, just doing philanthropic work such as building roads, schools and hospitals – social work – will not solve pain. When we feed the hungry, this is of course a good thing. But simply just feeding the hungry and nursing the sick won't overcome pain. Pain will not be finally overcome until we come to knowledge of the Self. When we know that my soul, I and God are One, then suffering will stop."

"Isn't it too easy to stay peaceful in your mountain cave, when there is nothing to bother you?" somebody asked.

The morning session ended with a response of impressive calm:

"Believe me – there are those who make a big display of throwing their money away and going off to live the life of a hermit. But if you are sitting in your cave and your head is filled only with ideas of how to get money out of those who come to visit you – how can that be peaceful? When you come to the Goal, there will be no difference between living in a house in New York and living in the silence of a Himalayan cave."

A short way beyond Laxman Jhula bridge stands the once-imposing travellers' hotel, the Bombay Guest House. I wandered within its doorway one afternoon, having heard that someone I knew might be staying there. He wasn't, but a recent police notice pinned to a wall caught my attention. The eyes staring directly at me from the badly-Xeroxed photograph were relaxed, alert, dark in an Asian way, their expression suggesting a seasoned traveller used to the ways of the country and to looking after himself. Beneath the photograph were the large black capitals:

MISSING
Jason Bellman. Aged 28. Dreadlocks. Eyes, brown.
Yin-Yang tattoo on right shoulder. When last seen,
wearing chappals and multi-coloured striped bedsheet.

Anyone with any information about the above person is
asked to contact Laxman Jhula police station, Rishikesh.

There was no mention of the man's nationality. Who had reported him missing? What had happened to him?

Some further light was shed on the mysterious disappearance later that very same day. The vehicle for this information was a Belgian – American whom I had last seen five years before in the Garhwal. Now I spotted him again, by the mail rack in Ved Nikatan. Around forty, he didn't seem to have changed at all.

"So you're back?" I said, as we exchanged greetings. "How long have you been here?"

"I never left," came the unexpected reply. "I'm at the *Bombay*."

His voice bore a hint of strain and still carried traces of early years in America. We sat down in a café for half an hour and the topic of the missing Westerner soon arose.

"I didn't know him; but I saw him at some of the yoga classes. He used to go for a swim in the river below the hotel every day...There's lots of theories. The current's very strong there; a whirlpool could easily have sucked him away. No-one saw him go."

"I heard one or two people got robbed over there," I said.

Dick, the man of at least three Motherlands took a short sip of tea.

"It's dangerous round Laxman Jhula after dark. I almost never walk about there alone at night."

"He smoked?"

"Yeah, it could have been to do with drugs. Some of those *babas*."

"You're suggesting they'd go as far as to kill someone!?"

"Probably they wouldn't. But they could've sold him some bad gear. He passed out; they panicked and pushed him in. All they found was his clothes folded up by the water. There's only so much the local police can do."

I shook my head. It was a congenial lifestyle for many travellers. One day melting into the next on the banks of the Ganga, sociable chai shops, cheap hotels and cheap drug deals. At night sitting round a fire, playing music, swapping anecdotes and rolling joints, with nothing to disturb the endless days except an occasional stomach disorder, until –

There was nothing further to add. Missing for more than two months, it was a long way to the Bay of Bengal where the Ganga emptied into the ocean. I pictured the bay. Sunrise breaking over a calm, pale sea; a line of nets raised on poles, waiting for the tide to carry the fish towards a strip of startlingly-white sand...

After a long pause: "Five years – here!?" For someone so sceptical of the ashram scene, it seemed like a feat of gross masochism to hang about so long.

His response made my attempt at levity seem frivolous:

"I'm sick of travelling," his voice faltered awkwardly. "Moving on – that's all I've known for the past fourteen years...I was four years at university in the States. Now..."

I could feel my own energy ebbing away as he spoke diffidently about this and that. The highlight of five years seemed to have been a three-week trek into the Zanskar range, north of Manali. There'd been an alertness about him last time...a kind of vibrancy. His spirit seemed greyer now, as though he'd lost it along the way. I tried to sound encouraging: "With your experience in the mountains, you could become a trekking guide – why not?"

He nodded as if it was a good idea. But I could see it didn't really grab him. The conversation stumbled to a halt. I was beginning to feel tired. When I'd finished what I was drinking, we parted company without arranging to meet again. He didn't say how he'd made a living during those fourteen years.

Death, whether it takes the form of a bloated, upside down cow by the roadside or corpses smouldering at the burning *ghats*, is always likely to hit you full between the eyebrows in India. When you least expect it.

Strolling back to the shade of my room one late afternoon I was surprised to see a small crowd gathered before an old sadhu's ramshackle hut, a dwelling which he had thrown up at the side of the path between the bazaar and the ashram. Whenever I passed, the bowed scarecrow figure would be sitting cross-legged within the ragged interior, a scriptural text propped up on a cloth bundle before him. Once his grizzled face lifted as I passed, but his eyes seemed to be gazing out upon another world. Like many before him and countless more who would come after, he had journeyed to Rishikesh because he knew that time was short. As I drew near I heard respectful, hushed voices. A man was issuing instructions in Hindi; another, detaching himself from the throng, hurried by me towards the bazaar.

"What's happened?" I asked the nearest bystander.

My informant gave me a solemn look:

"Holy man, he dead."

Another man with a better command of English enlightened me further:

"He has no family or relatives, sah. For nearly three months he has been unable to take solid food." He waved his hand at the bazaar. "Now he is dead, it is right that we give him proper burial – someone has been sent to buy cloth."

I lapsed into silence and decided to await events as inconspicuously as it is improbable for a Westerner in India. It was a shock to learn that the sadhu had died whilst I was out buying oranges.

There are no great cremation grounds at Rishikesh to compare with Varanasi, where the funeral pyres burn night and day, but the procedure for the most sacred ritual in the Hindu world is the same.

A frisson of uncertainty hung over the scene until the messenger returned carrying a length of the appropriate cloth. Some of the onlookers stepped aside to let him pass, giving me my first proper sight of the dead holy man. He had been laid full-length across the entrance to his little shelter, his shrunken limbs arranged in an attitude of repose. But it was his face which imprinted itself in my memory as no camera could. His expression was one of quiet rapture, of a man who had overcome the frailties of human existence. The eyes were closed but the mouth seemed about to stretch into a smile.

Although nobody knew him, the traditional rituals for the dead had to be performed by the living and the body suitably disposed of. Amongst Hindus, the sadhu is one of the few exceptions to the customary practice of cremating the dead on open funeral pyres. He is regarded as having already reached a state of purity in his evolution. It is only we lesser mortals whose souls need to be released through the transforming rite of the fire ceremony.

I watched as two men gently wrapped the white cloth about the old man, whilst the rest stood about in subdued respect. The slow murmur of ritual incantations began over the now shrouded form. As it was lifted on to a blanket, the silent watchful faces about me began to relax. The mood lightened; relieved smiles broke out: duty was almost done. The gathering parted as four men, each holding a corner of the blanket, raised the corpse to near-shoulder height and turned their eyes towards the waiting, onrushing waters of Mother Ganga.

Imagine the dignity attending a ceremony in which some king or queen is borne aloft on their last slow journey to burial – the attitude of those four pallbearers could have been no more pious had the unknown sadhu been one of their own family. A dryness rose in my throat as I watched them bear him away, a slow procession threading its way through the flat boulder field to the river's sandy edge. I was struck still by the transcendental simplicity of the spectacle, scarcely noticing that some of the spectators were already losing interest and beginning to drift away. A surreal mood, a commingling of joy and sadness held me to the spot.

As the figures reached the vicinity of the river, a distance of some seventy yards or so, their forms began to melt and dance in the water's reflected light. In no time at all, I could no longer distinguish one bearer from another, nor them from the frail burden which sagged between them. I could see little more than a wavering mirage of blueish light and darkness; spangles of diamond-hard light flashed deep into my eyes until I thought I was beginning to hallucinate. Blotches of liquid silhouette dissolved and reappeared in a shimmering display of jewel-like colours. It seemed like a dream, as though a thousand thousand fireflies cavorted about the river. For a moment, the figures of men reappeared, bending low thigh-deep in the swift current. Something – a package or a short pole, detached itself from them, bobbed once, then seemed to dart free. I saw it again, twice – a third time, then it vanished, swallowed forever by the kaleidoscope of dancing light.

I squinted into the sinking sun. But there was nothing more to see: the sadhu had gone. Only the tattered hut, a lonely bundle of clothing and his book of scriptures remained to indicate he had ever existed. I took a long breath and returned to the world.

On top of the high, whitewashed wall at my back a monkey scratched idly at its coat for fleas; a squatting beggar wrapped in an old blanket despite the heat raised his face through a pair of cheap spectacles. I felt the torrid heat again and remembered where I was bound.

Phantoms from my past flared briefly into my mind's vision, as I completed my short walk: uncles and aunties, school friends, a younger brother, and so on. J still lived in Sweden as far as I knew. The group on the ashram steps looked like a still from a 70s time warp. An African with six toes on his left foot was leaning against a pillar. He had lost his passport and slept in an empty hermit's cave a few hundred yards down river, or so he told those curious enough to ask – the idle remembrances flickered in my mind; and subsided.

A young Englishman dressed rickshaw *wallah* style in tight, sleeveless singlet and *lungi*, was staring back the way I had come:

"So the old guy kicked it?" He grinned in familiar fashion. Another with lank hair and lolling against the supportive arm of the concrete divan, exclaimed, "Yeah, that's right – they chucked him in the river."

I made no audible response and continued through the gate to my room.

The sadhu's hut remained undisturbed for a few days more. At certain times of day I thought it resembled a silvery chrysalis from which a rare butterfly had flown. Finally, early one morning, with the river running grey, I stumbled outside to find it gone.

For three weeks now, I had followed the ashram routine, arising at six every morning, performing various yoga *asanas*, meditating and before breakfast, paying my respects at the riverside. Most days I had devoted a few hours to exploring the immediate neighbourhood, visiting temples, following some unfamiliar lane or other, or sauntering along the peaceful banks of the Ganga itself. At first I tried to ignore the urge for change – after all, three weeks can scarcely be called commitment! But however colourful the local scene, I needed a short respite from the sacred city with its unusual odours and cheerful pilgrims jostling in the alleyways, not to mention the daily comings and goings of Westerners at the ashram, with their travellers' chat and hybrid perspectives. I tried to ignore the call, but it became increasingly insistent: and in the end I had to give way. The mountains, the sky, the fast-flowing river with its patiently waiting stones formed a web of knowledge – but how to tap into it? A trek towards the wilder regions up river seemed to offer an answer.

I paid *swamigi* a visit to inform him I would be away from the ashram for a few days. He broke into a spontaneous chuckle when I mentioned my need to move:

"Yes – it is very easy to get carried away when we read books written by spiritual masters or try to follow their instructions! But we should remember not to take their advice too literally. The hard rules, the renunciations – they were not originally intended for people of the world; those were intended for monks who had made special commitment to the spiritual path. So when these feelings come upon us – to wander for a while in the Himalayas, for example – we should not treat ourselves too harshly. The change will do us good. And when we return, we will bring fresh energy to our main purpose, spiritual growth."

I told him in parting about my incredible neighbour with his desire to be prime minister of his vast and complex country. The swami shook his head:

"This is a very strange man indeed - I have noticed him around the ashram. Indian people, they are much afraid of this man. For one thing, he is twice as tall as most of them." He laughed good-naturedly, "He wishes to rule India? Ha!"

My own plan was more straightforward. I would walk 50 miles up the Ganga to Devaprayag, a famous township which stands at the junction of the two rivers, Bhagirathi and Alaknanda, which meet there to form India's holiest river. Though quiet today, for many centuries it has been a staging point for pilgrims heading up the roadless valleys of those times to the mountain sanctuaries of Gangotri, Badrinath, and beyond. The town was also part of an ancient trading route to Tibet. In the early part of the twentieth century a man-eating leopard killed over 120 people in the district, before being tracked down and killed by the legendary Jim Corbett. Hopefully, I would not prove to be fodder for any descendants of this four-footed gourmet.

The map, which I had hurriedly copied from the wall of a restaurant, gave no hint of the ruggedness of the terrain between Rishikesh and Devaprayag. However, although the official trekking season was almost over, I thought snow unlikely. With regard to food and accommodation, I felt the best thing was just to set out and trust to fortune along the way.

Crossing a fragile section of the trail beyond Gangotri, with the cliffs of Chirbas Parbat above.

Shrine at Bhojbasa (3,792m). Bhagirathi
Parbat beyond (6,856m).

A holy man reads the sacred scriptures by the
Ganges, Rishikesh.

Hindu pilgrims leaving *puja*, Rishikesh.

Chirbasa (3,600m), en route for Gaumukh. Bhrigupanth (6,722m) in background.

Swami Brahmachidanya, the Gangotri yogi: "If I am happy here, why should I come to Japan, where nobody is happy!?"

Evening fire ceremony, Ram Jhula, Rishikesh.

Ved Nikatan ashram from the banks of the Ganges, Rishikesh.

The daily *satsang* within the ashram.

Laxman Jhula, the start of the author's trek upriver to Devaprayag.

The 'Cow's Mouth', source of the Ganges, Gaumukh (4,267m).

Early days. Author, centre, taking a break in the Himalayan foothills near Rishikesh.

Chapter 7

In the Footsteps of Laurel and Hardy

Before the days of the motorcar the Laxman Jhula suspension bridge was made of jute rope. Today's steel and wire cable structure stands only three miles up river from Rishikesh; and it was from there that my trek really began.

On the east bank, the crossing is overlooked by a curious, thirteen-storey temple which soars high above the river like some surreal, multi-layered wedding cake. As well as teaching yoga, spiritual knowledge and mystical science, the place also offers numerous nature-cures with well-qualified physicians in attendance to treat the varied needs of the droves of visitors who call upon their services. In the morning coolness before such throngs appeared, I found the vicinity a peaceful spot, where I liked to stroll or watch the early light shifting on the deep-flowing, Ganga. There was always a possibility, I had been told, of spotting an elephant or two come down to the river to drink.

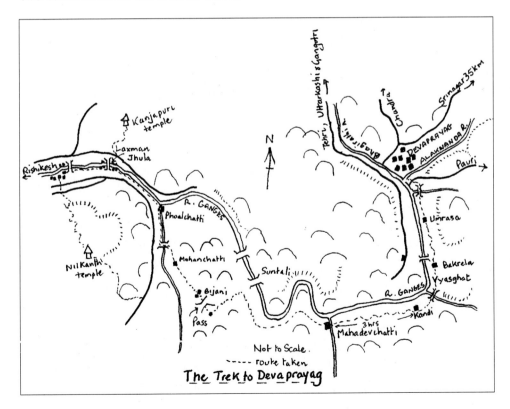

The Trek to Devaprayag

Not to Scale.
----- route taken

I set off past the temple, the energy of a journey's beginning coursing through my veins, my pack as light as a pillow and no worries about the future. Already, my short time by the Ganga had provided a precious glimpse of something of the truth in the poetic words of Jawaharlal Nehru:

"The Ganga, above all other rivers is still the river of India, which has held India's heart captive and drawn uncounted millions to her banks since the dawn of history."

I was grateful that the umpteen millions had chosen not to visit on the morning of my departure, for I had the winding dirt road to myself. I felt like a voyager setting out into uncharted territory. Even if it was only unknown to me, it was a special sensation. Occasionally, an overloaded Ambassador taxi trailing a funnel of swirling dust would dart past on its spiral back-route to Neelkanth temple which I had visited recently. Then the silence would descend again and I would hear only the faint rise and fall of my own breathing, or the slow creak from my rucksack straps. I might say at this point that I had had the opportunity to make the walk with two other Englishmen, but turned it down.

Steven, a long-haired hippy of diminutive height, informed me that he had arrived in India "with serious spiritual aspirations". All good intentions, however, evaporated when he ran into a six-foot four Londoner with a persistent line in laconic humour. They seemed an unlikely pair to strike up a friendship, but together they resembled a passable comedy duo. The impression quickly led Swami Dharmananda to dub them 'Laurel and Hardy'.

They must have got wind of my intentions, for the day before I was due to set out I was approached by "Hardy" who told me he and his friend wanted to walk up river for a few days:

"We're leaving this afternoon – do you want to join us?"

I had already overheard enough of their jocularity not to jump in feet first at the prospect of fifty mountain miles of English banter. However, I managed to keep these feelings to myself and simply explained that I preferred to do "this kind of walk" alone.

It felt good to leave the human turmoil behind me, if only for a few days. On either bank rose curtains of greens and powdery greys while here and there a projecting buttress caught my eye, lifting my gaze skywards to the silent hilltops. The scene was more than just a chaotic jumble of rocks and vegetation; it seemed a place rich in potentiality as well as history, as if it connected with an inner part of myself which I was yet only dimly aware.

Shortly, I passed two stout steel doors of recent construction set in the foot of a cliff face above the river. I thought they might hide tunnels and this surmise later proved correct when I was told they were part of the Tehri dam scheme, a controversial project a day's bus-ride up river. Construction was well under way, bringing the day closer when eighty thousand people would be uprooted from their homes. The dam would be the largest in Asia and would involve the sacrifice of twenty-three villages in the Tehri valley. Needless to say, the beneficiaries of the electricity generated would not be the local populace, but the residents and industries of big cities many miles distant. Opponents of the dam argue that because it is in an earthquake-prone region, the reservoir will be a disaster waiting to happen. Everywhere in India the visitor sees the sharp contrast between the timeless, mystical India of the imagination and guidebooks, and the scramble of a country trying to industrialise itself in the way of the West.

Half an hour beyond the tunnels the road swung right and I came to a hand written sign: <u>Phoalchatti</u>. It was the first place mentioned on my little map. There was a tranquil atmosphere, but I decided not to linger even though I had read that the most peaceful ashram in the area was at this spot – a tempting diversion which I might well have succumbed to had I encountered it two days later. A small boy eyeing me warily from the seat of a home-made bicycle and a young woman stooping in a field were the only signs of activity.

A little way past an empty chai stall, I left the road for a well-defined trail which carried me by a patchwork of cultivated gardens and a sprinkling of simple dwellings. So far I had been walking in the shade of the mountainside, but ahead I could see an opening to a broad, sun-filled valley. Before long a little bridge led me over a sidestream where I paused for a few moments to adjust the laces of my boots. Two or three miles later I reached the hamlet of Mohanchatti, 2¹/₂ hours from Laxman Jhula and a good place for a short break. The day's heat was already beginning to build up, reminding me of the need to avoid dehydration.

I swung my rucksack off my shoulders and took a seat in the solitary chai shop while the other occupants, two old men and the owner, looked on with mild curiosity. So far I hadn't sweated at all, which I hoped indicated I was in reasonable physical shape for the miles yet to come. Nevertheless, when the milky offering was set before me I was more ready for it than I realised. Under normal circumstances I do not drink sweetened tea, but I always found India's sweet, ubiquitous brew, with a hint of cardamon and cloves, far more refreshing than one of the numerous soft drinks which were also widely available.

After twenty minutes I was ready to move. I wasn't encouraged by the chai *wallah's* mournful countenance to make conversation, but thought I should try and asked whether two of my own countrymen had passed through the village. His expression instantly brightened. Yes, two "foreigners" had stopped at his restaurant the previous evening:

"One very tall, one very short."

I smiled to myself as I hoisted my pack aloft once more and stepped out into the street.

"Devaprayag?" I nodded into the deserted distance, noting to myself that it looked a stony trail from now on.

The chai *wallah* followed my gaze and seemed to read my thoughts:

"A hard road," he said, as if commiserating about life itself, as well as warning me not to take the foothills too much for granted.

At first the path wound at a gentle gradient along the left side of the valley which was already beginning to lose itself in a heat haze. It made no lasting impression upon me: there were no features of singular appearance which might impress themselves upon the memory. It was, I suppose, about half a mile broad, bowl-shaped and enclosed on all sides by foothills covered by a fuzzy, neutral-tinted down of vegetation. The path had led me temporarily away from the Ganges, which was hidden by the high ridge on my immediate left, an aspect which I hoped wouldn't be a recurring feature of the walk as I was already missing its companiable proximity.

Something else was puzzling too.

I had read enough to know that my route was an ancient Yatra trail followed by 'thousands' every year. But so far I had seen no-one. I had fondly imagined that every once in a while I would be exchanging salutes with penniless sadhus, though in fact the only person I had encountered on the trail from Phoalchatti was the small boy on the makeshift cycle. The height of the pilgrim season was some way past, so perhaps I was one of the last to set out before the weather changed and the snows barred the way to the mountain temples until the following spring. The day temperature offered no hint of imminent change, but it was the only explanation I could think of.

I was now no longer walking in the mountain's protective shadow and it wasn't long before I began to acknowledge the truth of the chai *wallah's* cheerful observation. A rocky, dusty undulation, the trail snaked unevenly through lifeless scrub and feebly irrigated fields. The sun was getting hotter by the hour and midday was only an hour away.

The jauntiness, with which I had set out, was already beginning to drain from my lower legs as my feet warmed up, whilst the temporary lift which the chai had given me was soon only a distant memory. Two more hours saw me nearing the end of the valley, at least the lower, flatter portion of it. The trail now rose more sharply towards a scattering of small dwellings and neat garden plots. At that point also, a curious thing happened – the trail vanished.

I glanced left and right thinking I must have missed a fork, but there was no sign of its continuation. According to my map it ought to be the village of Bijani, though I could only hope so. It was time to ask.

No sooner had the thought crossed my mind than three young children appeared from behind one of the houses; by means of laughter and arm waving they indicated I should climb the slope to join them. This I duly did, arriving at a baked terrace and the inviting shade of an open doorway. A young woman, very attractive, and probably their mother emerged from the darkness to greet me with a warm smile. They might have been expecting me so unaffected was their welcome.

When I requested *pani*, she motioned to me to sit down in the shade. She disappeared, to return a short while later with the water and accompanied by two more women about her own age. Word that a foreigner was in town had obviously spread rapidly. I took the glass gratefully as the new arrivals sat down, watching me shyly. To show interest, I made a couple of passing attempts to find out something of their lives, but received only giggles, more smiles and a refilled glass for my pains. Glad to be spared the usual battery of personal questions, I allowed my mind to sink into its own train of thought. In this mellow state, time slipped by and I could cheerfully have remained there for the rest of the afternoon. The situation was an agreeable one, but at the same time I had only eaten a handful of raisins in nearly five hours and was getting hungry. I decided the best thing would be to find a shady spot further on rather than impose myself on the villagers' hospitality any longer. I smiled my thanks and rose to go, dragging my rucksack with me into the harsh light of the open doorway. On the threshold I indicated a sprawling thicket at the rear of the little house and uttered yet again the powerful mantra – "Devaprayag, yes?"

The response was instantaneous. Two of the children and the youngest of the three women danced ahead of me up the loose slope, signalling that I should follow. Within a couple of minutes I suddenly found myself standing once more on the Yatra trail. Here, I paused to take a photograph of my three young helpers before waving goodbye. Their openness to a total stranger was touching.

I wondered, as I left the village slowly behind, whether other travellers had been confused by the footpath. Even with the benefit of a high vantage point overlooking the tilled fields, it was impossible to trace a line connecting the trail's lower end with its higher continuation. Later I discovered that Laurel and Hardy had blundered upon the hamlet in darkness, set the dog choirs barking and had been even more confused that I was. After floundering about for some time without a light, a door was eventually opened to them and there they spent the night.

The trail now rose steadily from the valley floor with thinning, desiccated vegetation on either hand. I was now totally exposed to the direct beat of the sun and felt vulnerable. As before, the way was deserted; the rasping croak of a solitary crow served only to emphasise my feeling of aloneness on the mountainside. But I was not as alone as I thought.

Rounding a corner with both thumbs by now hooked under my shoulder straps, I stopped short at the sight of a gang of men, Indian road-workers. Part of the trail had been completely washed away and the group, five or six in all, was busy making good the damage.

"Namaste!"

I stopped at the ragged edge of the repair work as two of the gang grinned down at me from their perch on an unstable-looking outcrop. One turbaned man in ancient, ex-army plimsolls, crouched in the meagre shade smoking a *beedi*; the rest worked barefoot amongst the loose stones. One, his semi-naked body gleaming with sweat, was plying the tip of a crowbar almost as tall as himself beneath a five hundred pound boulder, as yet still embedded in the rock face. It was one of those situations in which a few minutes' casual conversation would have been of mutual benefit. But whilst I had found little difficulty in communicating in India's towns and cities in my own language, I regretted not having even an elementary grounding in Hindi. I would have been interested to learn something of the trials of a Himalayan trail builder, whilst I'm sure the roadmen would have been only too willing to have a few moments idle talk with a wandering foreigner. Smiles, surreal pantomime and helpful hotel keepers will get you along when you want something, but there are obvious limitations.

The high sun was by now glaring out of a sky more white than blue and I was really beginning to need some solid food inside me. I narrowed my eyes against the reflected glare from the broken rubble and indicated to the man with the crowbar that I wished to proceed.

He understood me at once, shouted something to another man working higher up and waved me across the uneven gap with a wide grin. I don't know how old he was – forty or fifty-five – but he displayed the lean musculature of a much younger man. Without wishing to romanticise the labouring life, I couldn't help but compare their activity with the sedentary Western life which forces millions of both sexes to spend much time and money working out in gyms or jogging to achieve the same result. Giving the loosened boulder above me a wary look, I scrambled over the unconsolidated stone bed to secure ground on the other side. I glanced back as I walked on, but the crowbar man's attention was now back to his task.

Twenty minutes later when the trail dipped into a shallow wooded ravine I heaved a sigh of relief.

The ravine was little more than an indentation in the hillside, but it provided the shade I was seeking. I flung off my rucksack and slumped down next to it with my back against a tree. The only thing lacking was the gurgle of water in the dried-up stream bed at my feet. But my trusty wineskin was still half-full and I splashed some of its tepid contents over my face and neck, before pouring two or three gulps down my dry throat. I'd noticed over the last hour that I had stopped sweating and knew it was essential for me to maintain a regular fluid intake. The water made me feel much better and my attention turned to one of the two side-pockets of my dusty, much travelled rucksack. Lunch, hopefully, was not too battered by the occasional knock it might have received along the way.

I exhumed each item one by one and set them on the small plastic plate which I fished out from the main compartment: a tomato and cucumber sandwich, three or four glucose biscuits, a chunk of peanut brittle, an apple and a slice from the pineapple I had taken the trouble to carry along. A mood of quiet repose soon fell upon me and I recalled, as I ate, a couple of other occasions when the heat had been more of an adversary than a blessing.

One summer I was in Morocco. The temperature was around forty-five degrees centigrade at Zagora, in former times a camel-route outpost town on the edge of the desert. Today not much happens. I got caught with an empty water bottle.

"Could I," I remember thinking, "reach that next patch of shade two hundred yards across open ground, without passing out?"

I did of course but the experience was a salutary one which left me in no doubt as to the possible consequences of being without water in such temperatures.

Shortly, after I had eaten and kicked off my boots and socks to bring some relief to my hot feet, the inevitable happened, and I drifted into a shallow doze. Mountain travel may have much to do with beards, sweat and punishment but I am glad to put on record that I took my hour's siesta without flinching. Upon awakening I peered through a gap in the withered branches at the way ahead: after long hours of repetitive footfall, it was all that concerned me. The hilltops were a faint, tenuous line, as if sketched on tissue paper. Nearer at hand nothing stirred; not a single insect moved, nor did anyone pass by. There was just me, a fleeting patch of shade, and the journey.

And the pair ahead, Laurel and Hardy – what of them? Considering how quick I had been to reject their company, it was a little odd that they should continue to pop up in my mind. In the heat's shimmering haze I imagined seeing them at the next bend...or the one after that. The predominant feature marked on my map after Bijani was a pass, but since no scale was indicated, I could only guess at the distance I had still to face. I checked my watch, it was already nearly three in the afternoon. With a final gulp of water, I gathered my scattered belongings together and once more settled into a regular stride.

I had set what I considered to be a good pace since leaving Laxman Jhula, but now the notion of getting to the top of the pass, some eighteen hundred feet above me, dominated my mind. I could not see it at all – there was no telltale notch on the skyline so far – but a sequence of empty, half-mile zig-zags up the mountain-side dauntingly pointed the way. Somewhere up there, I hoped would be refuge, a shelter of some kind, often called a 'dharamsala', which provides basic overnight accommodation for those who wander the local trails of these regions. I had thought of carrying a collapsible umbrella with me to use as a sunshade, but on a grey day back in England, I had abandoned that idea as an unnecessary luxury. Now I was beginning to wish I hadn't. However, there were still a number of factors in my favour and it was better to count my blessings. Its contents were a trifle tepid but my water container was still half-full, a pint or so of liquid remained. My feet, though not beautiful to behold were so far without blisters or sore spots; the route-finding was not unduly taxing, and last but not least, my enthusiasm was in no way diminished, despite the blazing heat. The main concern was to avoid further dehydration by attempting to rush the last hours of the day.

I tramped on, a steepening drop opened up on my right into the valley and the terrain became more arid the higher I climbed. Glimpses of sage-coloured vegetation far below, a bone-dry watercourse above the path and the sun burning through the shoulders of my cotton T-shirt, prompted a stream of cooling imagery to flow through my mind. In one such image I saw myself sitting under a village handpump as its crystal waters cascaded over my head and shoulders. I tried to banish such fantasies for they did me no good, and limited myself to a sip or two from my water container every twenty minutes. An unusual but simple method of yoga breathing through the tongue folded into a tube across my mouth, was also helpful. The air which passes into the throat in this way resembles a drink. It is good to do this when morale begins to flag in hot and trying conditions.

By regulating the breath the mind gradually becomes focused in the present: the here and now. I inhaled to the count of three; exhaled to the count of three or four (the ratio depends upon the gradient). Eventually my respiration and stride became synchronised in a single flow of steady movement, no beginning, no end. One begins to walk with the regularity of a pendulum with no jerkiness. Anyone approaching me and hearing the loud hissing note so produced, would probably wish to give me a wide berth, but its effect is one which I have regularly found helpful. Afterwards the air is expelled through the nostrils, and the exercise is repeated for as many times as is felt necessary.

In some strange way it also prevented my mind from growing dull and fixing itself upon trivialities. The trail, even the burning slopes around me, seemed to pulse faintly as if in response to the rhythm of my walking. When I asked, "Who is walking?" "Who is it feeling these discomforts?" my mind slowly switched into another mode. I began to experience these things in a less *personal* way. Persistence was clearly worthwhile for it seemed to put me in touch with a higher part of myself, a desireless, always constant state which simply accepted conditions instead of fighting them. It felt like protection against external things. Only when I allowed my mind to wander or stubbed my toe against a half-buried rock did I lose it.

I kept going in this way for an hour before stopping to tighten one of my laces which had come undone. When I stood up my head unexpectedly reeled and for a few moments I felt quite giddy. I glanced about me but there was still no shade unless you were a snake or a rodent. The steep shale slope on my right gave my back a temporary support whilst I allowed myself another drink. I now had a bird's eye view of the deep side valley up which I was now trudging, one raking incline after another followed the folded indentations of the mountainside, with next to nothing to catch the eye in the way of landmarks. The road menders and the wider valley where I had been kindly received by the villagers, were lost to view in a khaki haze. I looked hard for signs of life on the gritty ribbon of trail, but the way ahead and the trail behind were empty. On one occasion I fixed an unblinking stare upon a small dark speck for some time, convinced it was another traveller resting at the foot of the shale bank. At any moment it would start to crawl after me like a tiny stick insect. I dragged my eyes away at last. Laurel and Hardy, if they were still going, would probably have done this stretch of the route in the cool hours of early morning. Lucky them!

A grey lizard sunbathing on a slab of greyish rock inspected me, did not budge.

Alone, it was all-too easy to exaggerate the difficulty – or, for that matter, underestimate the obstacles. Would it be one more hour? Or two? I wasn't sure. Since I was carrying a lighter and a small cooking pot I could easily have stopped there and then and camped for the night without a tent. On the other hand, I was longing to see the Ganga again and dropped the idea almost as soon as it arose. For a few seconds I allowed my breath to settle into its former rhythm before pushing myself away from the slope. I tramped two or three more bends before the road's angle began to ease until, finally, I reached the top of the pass at about five-thirty.

It was little more than a depression in a bare rock ridge, but my legs received new energy as I spotted a simple refuge or small cabin, standing at the base of a fifteen-foot abutment. An agreeable aroma wafted from the hut as I approached the open doorway, where a man was crouching amongst a clutter of blackened pots and kettles.

"*Namaste.*" My voice had the scratchiness of a second-hand tape.

The man's eyes shone brightly and appraised me with a questioning look: "Chai?"

I nodded gratefully and a few moments later, a glass of the steaming liquid was handed out to me. I gulped it down in a series of scalding sips, requested another, and sank on to the solitary bench with my back against the hut's stone and timber wall. Standing a little to one side, a well-dressed Indian couple in their early thirties were also taking refreshment and admiring the view. The woman was dressed traditionally in a sari of greens and yellows. Unlikely in such craggy surroundings, the man was wearing an ill-fitting Indian version of a Western city suit in powder blue and, even more incongruously, a fluffy pink balaclava, which made me think of Egyptian mummies. They both smiled warmly as I looked in their direction, but my voice was too lost in my throat for me to attempt conversation after so many hours on my own. Where they had come from it was impossible to tell, though their footwear spoke more of the pavements of Delhi than the rough trails of the Himalayas.

My gaze roamed past them and the stony terrace. The harsh glare of the valley I had just left was gone now and a mellower light was beginning to bathe the heights. The ground dropped steeply from the ledge, a canopy of deep-green forest leading my eye down over two thousand feet to the ravine of the Ganga. Except for one place, a break in the precipice where a flood of light glinted on the water, the gorge was already deep in shadow. Yet it was a comfort to know that we were neighbours once again. A few minutes later the Indian couple strolled away after depositing their glasses with the tea-*wallah*. I waited a while longer before deciding to push on a little further, myself. Dusk was gently advancing over the orange and rose-tinted crests and it seemed an ideal evening for sleeping under the stars. However, the unexpected hadn't quite finished with me yet.

The trail had obviously been a major enterprise at some time in the past, for it was a succession of level hairpin bends hewn from the mountainside and on which no plant seemed to grow. In less than half an hour it brought me to a painted Hindi sign pointing steeply down into the forest. A quick check with my map suggested this was very likely to be the turn-off to Suntali, another small village. There was little point in starting the descent so late in the evening, so I decided I would spend the night at this spot. I gathered some twigs and dry grasses from the nearest reaches of the forest and soon had the satisfaction of seeing half a pint of soup bubbling over a little fire. Strangely, I was not particularly hungry at that point and made do with the soup and a couple of rough-cut slices of squashed bread. By the time I finished, the dying sun was already preparing to dip behind the golden tips of the most distant hill tops; night was imperceptibly falling and in another thirty minutes the battlement-like ridge above the trail would be in darkness. In this serene and profound silence I began to unroll my sleeping bag, having first cleared a suitable area of any unwelcome stones. Sleep, however, was not to be mine for a while yet. Hardly had I pulled my sleeping bag up to my armpits, when I heard the sound of approaching voices.

At first I could not tell from which direction they were coming, but a few seconds later several young women appeared from below, each bearing a large bundle of dry sticks. There was a moment's lull in the chatter as I was spotted, lying like some large blue bug impeding their progress. Amused glances were thrown upon me as they trod lightly past my recumbent form in single file, before slowly, and with unconscious grace, they disappeared one by one into the deepening stillness of the Himalayan nightfall.

Five minutes passed.

Then, just as I was deliberating whether it had been a good idea to turn-in wearing trousers I heard the snap of a branch in the darkness behind me. I sat bolt upright as a couple of stealthy footfalls followed. An unseen threat is often more disturbing than one which is plainly visible and I waited with no little concern to see what would happen next. My boots were close at hand whilst my pack had been propping my head as a pillow. A form moved within the indigo-blue of the forest and next moment a man emerged, an old man in a turban, grey-bearded and carrying a long and very ancient rifle slung over one shoulder. Upon seeing me, he stopped abruptly, fixed me with a fierce stare and began to talk rapidly in Hindi, all the while waving his rifle this way and that in a very agitated manner. The surreal notion popped into my mind that this was the landowner come to demand a camping fee. Watching the man's antics closely, however, brought the slow realisation that he wasn't about to rob me at gunpoint: he was trying to warn me of something - but what?

A three-quarter moon was already edging the rock battlements with a luminous braid, so a storm in the immediate future seemed most unlikely. I could only mutter a few words in reply and point with assumed confidence at the gap in the foliage through which he had just stepped:

"Me, morningtime," I said, making an absurd snoring noise in my nasal passages in an attempt to clarify the situation.

At last the man shrugged and seemed to decide he was getting nowhere: with a final flourish of Hindi, he re-shouldered his battered gun and began to walk away in the same direction the firewood gatherers had taken. A little taken aback, I watched his form bob slowly away into the shadows until he too disappeared altogether. An overpowering silence enveloped me once more. Although I was physically tired, my mind was now very much wide-awake. I struggled from the confines of my sleeping bag and lit a thoughtful *beedi*, the pinpoint glow in the darkness giving me some little comfort. But what to do now? For the first time my mind began to turn on the nocturnal habits of snakes, scorpions and other

135

inhabitants of the curfew hours. A peculiar shriek such as I had never heard before suddenly rang out from far down in the inky forest, as if to point out the isolation of my own situation.

When I reached the end of the *beedi*, I stubbed it in the dirt and, without any clear reason why, began to pull on my boots. At that moment I suddenly began to feel a slow prickling sensation that I was being watched.

I glanced up sharply and there, a mere twenty feet away, a small boy holding a white horse by a rope, stood watching me. How long he had been there I had no idea, but he continued to direct a look of concerned curiosity upon me. I don't know who was the more surprised. After a pause I found my voice and, directing a finger at myself, explained,

"Me sleeping…Here. Morningtime Suntali – Yes? Breakfast – chai, Suntali…Tomorrow. Okay?" I was beginning to get a little agitated at being disturbed so many times.

At this, the boy's eyes widened; and, to my surprise, he began to address me in English:

"You not sleep here. This very bad place. Four men die here. There are ghosts…animals. This is not good place for you!"

"You have seen?"

The boy nodded urgently:

"I have seen ghosts of men walking about. It is better that you go back to *dharamsala* – only a few minutes…Sleep the night there."

His eyes glowed in the near-invisibility of his face and were full of earnest conviction. Had the old man been trying to tell me the same thing?

I nodded to indicate I understood what he was telling me. The boy lingered for a few minutes more until, perhaps feeling he had discharged his duty, he turned and began to lead the horse away. Both he and the silent beast seemed to drift off into the darkness, lit by a curious glow of light, which I assumed must by caused by the rising moon. Now the atmosphere of the place was no longer as attractive as before.

The ridge and a number of fallen boulders seemed to pulse with a strange phosphorescence against the encompassing blackness. All of a sudden it was a very lonely place indeed. The boy was right. Even if there were no spirits of the dead to haunt me in my sleep, I would probably have a more restful night in a proper bed. With a bit of luck, there would be other starry nights after this one. A second harsh screech from the blackness below and my mind was made up. However, I had back-tracked no more than a few yards when a hair-tingling thought stopped me dead. I suddenly realised that as the horse was being led away, I had heard no sound from its hooves on the rocky ground! I peered into the darkness behind me: there was nothing there.

136

By the time I came once more to the vicinity of the *dharamsala*, the moon was showing me phantasmagorical shapes amongst the rocks and I felt glad to be in the vicinity of human habitation again.

The Victorian writer and traveller, Robert Louis Stevenson once proclaimed, "I travel not to go anywhere, but to go". More than once I have echoed such lofty sentiments too. But there must have been days when he and his donkey were glad when the day's journey was over. When I saw the pale lantern glow shining its welcome from the door of the simple hut, I know I was.

A quiet murmur of voices greeted me as I angled my head beneath the low portal. The Indian couple had gone, but the chai-*wallah* had been joined by a companion, another man who sat just within the circle of light cast by the lamp. I had arrived at precisely the right time for a meal of rice, *dal* and chapatis was about to be dished up.

My appetite, dulled by the hours of walking, had meanwhile returned and I needed no second bidding when the lodge-keeper motioned me to the floor. Hardly had I sat down when a full plate was handed to me and I set about it like a starving man. It is surprising how even the most ordinary food can seem attractive and flavoursome under such circumstances. Three glasses of chai helped to wash down the meal which included five or six chapattis. Almost as soon as I finished the first plate, another daunting wallop of rice was offered me, but I made indications that I was full and declined it with a shake of the head. A feeling of great drowsiness quickly overtook me and sleep, as far as I was concerned, was the next item on the menu.

The *wallah* spoke no English, but a suitable hand-pantomime produced an immediate wave towards the pitch-dark sanctuary of an inner room. It was just as well the *Yatra* trail was deserted, I reflected thankfully, groping my way into the gloom before coming hard up against the item I was looking for. The *charpoy* was a typical Indian rope bed, seen in every village, often standing outside a dwelling with some prostrate male lying upon it. I was curious to try it out.

I lit a candle stub and looked around.

There was no window and the bed, some five feet across, took up a disproportionate amount of space. The rest of the chamber was occupied by a stack of firewood, a jumbled disorder of rusting ironwear – and a horse. Sharing your bedroom with a horse isn't everybody's idea of luxury travel, but I was too sleepy to care about such trifles. Tethered to a post in the corner furthest from the foot of the bed, the animal was head deep in a Hessian feedbag. Its ears pricked up as the candle spluttered into life and it glanced round at me with an expression of mild curiosity in its eyes. So long as it kept to its own territory I had no qualms.

I undressed and crawled into my own sack, where for a few minutes I lay awake savouring the still night, the strenuous events of the day and the low murmuring of voices from the other room. By the outer door, three mules periodically stamped their feet on the hard ground with a musical jingling of bells. But shortly I was soon unconscious of even these innocent distractions as I dropped into a deep sleep. I awoke only once during the night – to hear a scuffling noise by my rucksack which I'd left at the foot of the bed. One of my boots hurled with instinctive accuracy towards the disturbance produced the desired response.

Dawn came too early for me to think of anything more than a perfunctory lie-in. Light streamed across the threshold of the outer door and unobtrusive movements came from the adjacent room. The horse had shown itself to be a considerate stable companion and I felt much refreshed by the seven hours rest.

The two men, barely moving, crouched by the *chulha*, or earth fireplace, as though they had never left it. As I appeared, the *wallah* tossed a tumbler of milk into the already bubbling pan and followed it with a twitch of spice. I shook my head at his offer of the previous night's cold leftovers, but was more than ready to start the day with a tea. Two glasses later I was ready to make my farewells and be on my way.

"*Kitna Hai?*" I asked, offering *beedis*.

"Ten rupees," the *wallah* responded calmly.

So he knew <u>some</u> English – I shouldn't have been surprised. Ten rupees was little more than the cost of a postcard stamp.

As soon as I hoisted my pack on my shoulders, they made it clear that twelve hours walking was a long first day out. I smiled my thanks at the door and checked my watch. It was a few minutes past seven. As I set off I wondered what the likes of 'Nixon' would have made of such a place.

The tilting rays of the early morning sun revealed a magical world of forested valleys, hill tops and bare orange rock; far below the Ganga wore a sheen like mother of pearl: it was a sight to inspire even the most stolid imagination. Keeping close to the edge of the mule trail for a better view of the drop I soon reached the scene of my strange encounters of the night before. It was deserted now and, seeing the place robbed of much of its mystery by daylight, the commonsense side of my nature began to wonder if the events amounted to no more than projections of my own imagination. If there were malignant forces hovering about that rocky lookout, I was in too positive a mood to feel threatened by them. Reason has a wonderful way of denying the validity of anything which is beyond its powers to explain, but personally, I wouldn't be in too much of a hurry to deny that the spirits of dead road-builders might have lingered about the spot where they met their end.

I paused for a moment or two at the Sintali sign. It was truly an exhilarating viewpoint, full of intense colours and clarity of form. I felt sorry to be descending so soon from the heights, but in a few hours time I might be grateful for the protection of the trees. Far below was the gleam of water.

The path took me steeply down the forested mountain-side and it was evidently much-used, judging by its compacted surface and the polished gloss of many embedded stones. I fancied that more pilgrims walked to the holy places in former days out of necessity, whereas today, the bus-riding tourist-pilgrim is a relatively new phenomenon. More affluent than their parents and grandparents ever were, these Indians journey from all over the subcontinent during the four months or so in the summer when the Himalayan roads are free of snow. The grimy, low-geared buses as well as the taxis, do a brisk business, whilst the dharamsala where I had just spent the night was only one of many such lodges to be found every few miles.

"Why walk if the bus can take you there in four or five hours?" somebody may ask. Usually, people who ask those kind of questions don't have much empathy for the kind of answers one might give. Nowadays, I don't try too hard to convince such people that it would be right for them too.

The path took me steadily downhill, the views soon being restricted by the tree cover, but I had occasional glimpses of the Ganga getting steadily closer, or of some little hamlet, or *chatti*, jutting above the trees. Now and again villagers passed me singly or in groups, making their way quietly up the trail in search of firewood. After two hours of a steady, traversing descent I once again found myself in proximity to the river; it was now only a stone's throw to my left. Rising above the trees directly ahead were the substantial twin-posts of the Suntali suspension bridge, but of the village itself, I could see no sign. I thought I must have missed a turn-off as my pace picked up a gear on the easier gradient. I had not so far stopped or eaten since my early morning cuppa and had promised myself a break when I reached the village.

Twenty minutes passed before the trail's meandering course brought me to the crossing. The map at this point was unclear – or I had caused the confusion myself by copying it down too hastily into my notebook. Did the bridge span a sizeable tributary or the Ganga itself? Puzzled over the situation and feeling irritated with myself for the careless oversight, I inspected the 'tributary' more closely. Surely it was much too bold a stretch of water, too full and self-conscious in its surging power, to be a mere tributary? Eventually the penny dropped and I realised I didn't have to cross the bridge just because it happened to be there! The 'tributary' was the Ganga and I should stay on my side of the river. It swung left, away from the direction I thought – or hoped – it would be heading and disappeared round the foot of a vegetated cliff several hundred feet in height.

I really wanted to stay close to the river and was sorely tempted to try and pick my way across the obstacle: steep though it was, it took only a second or two for my eye to create a route linking one patch of drab, steep grey rock with the next. Much of the striated slabs, however, were draped with deceptively innocuous greenery, and previous experience told me that, like people, such places were usually much more complicated close up.

I decided not to chance it.

To the right of the bridge and some seven hundred feet above the spot where I now stood, was a small but distinctive U-shaped col. It looked like the entrance to a hanging valley. The glint from a sloping roof made me wonder if I was not already closer to the hamlet of Mahadevchatti than I realised. If I was right, it would be a good excuse for a proper breakfast stop when I got there. The temperature was already rising and I took a couple of pulls from my wineskin before starting the ascent.

As soon as I began to climb the exertions of the previous day began to make themselves felt in various ways; my legs had lost much of their spring, a condition which I hoped might be due more to skipping breakfast than tiredness. The path was steeper than it looked from below and my pace up the winding diagonals of the mountainside gradually dwindled to a crawl. The col was a long time in coming and after half an hour I paused to scan the scene below.

It was certainly a picturesque spot. In 1929, Mahatma Gandhi, taking a break from his struggle for his country's freedom made a visit to the Garhwal region. With his scriptural simplicity of language, he wrote:

"In these hills, Nature's hospitality eclipses all man can ever do. The enchanting beauties of the Himalayas, their bracing climate and the soothing green that envelop you leaves nothing more to be desired…Peace, which is the final end for the soul dwells in these shrines of Garhwal."

Enclosed between the hot Gangetic plains and the high frozen plateau of Tibet, few Westerners had probed its hidden recesses until well into the twentieth century. The difficulties of approach, the scarcity of all but the humblest of facilities, the absence of modern transport and the need for a special permit for border-sensitive districts deterred all but the most persistent of travellers. Compared with the hospitable facilities to be found in the major trekking regions of Nepal, the Indian Himalayas are relatively undeveloped. The lone traveller needs to be prepared to put up for the night in unlikely places. One can carry a tent, but it is all extra weight and only serves to further insulate one from the hill communities amongst whom one travels. Given half-decent weather, one can apply a tent-free approach even to the lower glacier regions, as the later pages will describe.

A modest amount of persistence brought me at last to the col, a bare place, thinly wooded and boasting the decaying walls of a temple, as well as a couple of forlorn huts. The rest of its charms, if indeed it had any, were kept well hidden. Thankfully, it had one redeeming feature - a small chai shop at which I quickly discovered that Mahadevchatti was still ahead of me.

While I waited for a tea, I ran my hungry eye over the shop's meagre food display.

If mine was an ascetic's journey of frugal self-denial, then I had stopped at the right place, for the shop had only one item for sale: glucose biscuits to be precise, a dozen packets strewn across the grubby glass shelf. A chubby three-year old child beamed at me from each wrapper, a brand which I recognised as the worst I had ever tasted. Their soapy flavour might have been forgiven had their energy content been of some value, but there was probably more glucose in the wrapping than in the actual biscuits. Beggars could not be too fussy in the circumstances, however, and I bought two packets.

I also decided that now was a good time to visit the expedition chiropodist.

A rapid inspection established that the only cause for slight concern was my left foot: a small sore had appeared on the top of my second toe and a patch of skin on the ball of the same foot was sensitive to fingertip pressure. The toe, a slightly enlarged legacy of 'character building' at boarding school, was always the first to protest if ever I wore boots a shade too small. In hot countries paying close attention to such warning signs is the key to maintaining health, so I wrapped a strip of protective plaster about the toe after first cleaning it with a swab of cotton wool dipped in water. Cleanliness of even the smallest of cuts is very important, because if left untreated they are likely to deteriorate and turn septic very rapidly. On the other hand, the body as an organism, has astonishing self-healing powers of its own. With diarrhea for example, the right medicine is often no medicine.

Lack of space prevents me from dwelling at length on how modern and traditional medicine operate side by side in India. But when a person falls sick a wide choice of treatments is available. There are Western-style doctors schooled in modern treatment or symptoms; *vaids*, who are trained in the holistic principles of Ayurveda, one of the most ancient systems of medicine in the world; *hakims*, schooled in Greek and Arab practices which employ herbal pastes made from plants, ground-up minerals and precious stones. There are numerous variants of the last two, both of which take account of the role of mental health in maintaining physical 'balance'. Ayurveda is a mixture of science, art and philosophy. Rest and fast is often the foremost therapy, The *vaids* derive their remedies from herbs and potions prescribed in ancient texts such as the Vedas.

Among other things in my compact medical kit, I was carrying a small tub of

Calendula ointment, a herbal remedy for slow-healing cuts and a self-mixed rehydration powder comprising one part salt, one part bicarbonate of soda and eight parts glucose powder. To maintain my general health I was taking two garlic perles each morning. I stopped taking anti-malaria pills in northern India after Dharmananda told me there hadn't been a single recorded case of malaria in Rishikesh for more than two years.

For more subtle forms of 'dis-ease' such as feelings of negativity when travelling, a different approach may be needed. Again it is the breath which can bring well-being.

All I do when unhelpful states arise is focus on it, the breath. I breathe in slowly, consciously – and deeply. If I find it a little difficult to expand the lungs sufficiently whilst walking with a rucksack, it maybe better to stop for a while. People think that shallow breathing is normal. It isn't. If you practise it often enough, eventually you start to breathe deeply without having to think about it. I see myself drawing in energy, strength, calmness – or whichever quality I need at that moment. I happen to know the energy as '*prana*', whilst others may know it as '*chi*', or 'cosmic energy' or whatever. The *prana* can be experienced as light, filling not only the lungs, but the entire body up to the point between the eyebrows. Keep up the conscious breathing for as long as it takes which isn't usually long. The unhelpful states begin to flow out of the body with each exhalation. I know when its working because a change begins to come over me. Soon I feel all right again. Of course I'm not pretending that simply 'feeling good' today, through practising strange-seeming techniques is a kind of spiritual transformation. Some people might, but it's only a step on the path. If you're always filled with a great sense of urgency when travelling, of thinking you have to get somewhere <u>fast</u>, then you could consider the advantages of a spell of <u>conscious breathing</u>, of slowing down. Anyway, that's what I try to practise.

The morning was slowly advancing and after a second chai and a packet of 'baby' biscuits I was ready to leave.

Thankful both for the break and to be away from the oppressive air of the col, I pushed on at a steady pace along the trail's undulating traverse of the mountainside. Before long the overhead foliage began to thin and I received regular reminders of the previous day's experiences under the direct rays of the sun.

Once, I glanced through a 'window' in the fretwork of branches to see a towering precipice on the opposite side of the river. I stopped to look as a movement caught my eye. Far, far above me on the immense rock face drained of all colour, a black speck was crawling slowly like some tiny Lilliputian insect. Behind it a gossamer-like thread unwound itself steeply downwards and out of sight round another buttress of dizzying height. It was the road to Devaprayag. The contrast between that immensely high mountain wall and my own puny self was a humbling experience. I shuddered to think of the certain fate that awaited a driver's slightest miscalculation 1500 feet above the Ganga.

Mahadevchatti, which I reached in less than an hour's steady walking, provided the usual simple amenities including shade. But I resisted the call of the chai-kettle for once because I intended to make a longer halt further on. According to my map, I should reach a bridge spanning an unnamed tributary within the next three miles. I strode on, with only a single pause for a few raisins and a mouthful of water to interrupt my stride.

By now I had put nearly thirty miles between me and Ved Nikatan, which meant I was over halfway to Devaprayag. It wasn't so far in distance, yet the ashram and its activities already felt an age away. Moreover, I had seen no sign of other foreigners with whom I might have passed a companiable hour. Signposts, too, were also prominent by their absence. Yet so long as I kept the Ganga within throwing distance on my left, I knew I couldn't go far wrong on my outing through Uttarakhand, as the two major hill regions Kumaon and Garhwal, are collectively known. On the odd occasion when I did feel unsure of the way, I simply checked with the next villager who crossed my path. The response to a stranger in their midst was rarely less than helpful and friendly. Though I wandered alone through a strange land, I had no such feelings of the kind of loneliness I have on occasions felt when chance has dumped me in some unfamiliar concrete city in the West. Now and again, especially when my mind was free from internal chatter, I would feel a powerful connection with some insignificant place, as if I had known it at another time.

I have mentioned rebirth in an earlier episode, but I think the subject is worth more discussion, if only a brief one. How helpful is it, for example, for a person to know of a previous existence which he may or may not have led in a past life? Does it help to illuminate this one? Once, I was told by a clairvoyant, that in a previous life I had been a monk who failed to make full use of his powers and had been reborn in order to complete unfinished tasks. Perhaps. It is impossible to prove either way. But it is difficult to see how the tragic episodes in human life can be satisfactorily explained otherwise. Rebirth might explain why I have been drawn again and again to solitary places, whether it be the Himalayas or a standing stone by a lonely Scottish loch. Is the inclination a carry-over from another life? The questions will never end if one seeks that kind of proof.

It is true that the notion of rebirth offers the opportunity for self-delusion on a grand scale. To mention only a few encounters: I once met a woman who claimed to have been a priest in the Lost World of Atlantis; also a man who assured me he had in former times been a Tibetan despot who did away with people simply by thinking about them. Whatever the truth of these claims, the human pilgrimage goes on. And always will.

In *The Journey to the East*, Hermann Hesse writes of the eternal human quest:
"this procession of believers and disciples had always and incessantly been moving towards the East, towards the home of Light. Throughout the centuries it had been on the way, towards light and wonder, and each member, each group, indeed our whole host and its great pilgrimage, was only a wave in the eternal stream of human beings, of the eternal strivings of the human spirit towards the East, towards Home."

The counter arguments contending that such 'beliefs' are illusory, draw their ammunition from heredity and environmental evidence to explain why we are who we are. An enormous topic and I have no wish to write a treatise on the subject here, even if I was capable of it. Suffice to say that just because a materialist denies the existence of rebirth doesn't extinguish the evidence for it. Perhaps we should follow the wisdom of the Middle Way: that what we are now and will become in the future is more important than what we once were? Otherwise the past will constantly haunt our lives.

The path now wound across the wooded hillside, sometimes dropping two or three hundred feet before climbing again. Although it was obviously a very much used trail of great age, I scarcely saw anyone. Most of the way the trees shut me off from a sight of the Ganga, but the restricted view only made the unexpected glimpses of the river more memorable when they came. I plodded on for another forty-five minutes. The long shadows of early morning had by now shrunk to almost nothing as the sun climbed towards midday. Whilst I had been warned about the dangers of trekking alone in the Himalayas, I knew I was experienced enough to be confident I could deal with such terrain. It wasn't after all the frozen tablelands of northern Tibet. Even so, the day was to end in a way I had not anticipated.

In a short while the trail began to dip again towards the river, the view suddenly opened up and I saw I was within three hundred yards of an imposing cliff I had spotted earlier. A sparkling movement near its foot made me quicken my stride at the thought of a stream. I halted within the shade of the vegetated rock face, threw off my rucksack and ducked my head under the modest water spout which sprang conveniently from a mossy recess at shoulder height. Someone had created a funnel-effect with a broad leaf so as to increase the force of the dropping water: I had not had a proper wash all day and gratefully gave my head, neck and arms a good soaking. The water was much colder than what I had been drinking and its bubbling energy stimulated my whole system. However, my ash-coloured boots told their own story and when, no more than a hundred paces later, a dry streambed appeared, I decided to have a longer break. It was now the hottest part of the day.

I boiled half a pint of tea water over a twig fire amongst a jumble of protective boulders and ate some fruit, tossing the core in the direction of some birds chirping amongst the nearby undergrowth. For the next two hours I snoozed away on my back, having first gathered my scattered possessions into a less conspicuous bundle at arm's reach. Though I wasn't particularly afraid of theft, it seemed a prudent measure not to lay temptation before passing eyes.

After my siesta I trudged on, sluggishly at first, but with a slowly clearing mind. The sun's glare had softened somewhat by now, though the riverside rocks still shone with a metallic gleam. Once or twice I felt the sharp pressure of a flinty stone through the sole of my boot, which acted as a mild warning to take care of my feet.

I have mentioned I was alone on the trail, but this was not entirely true, since I was accompanied by the constant twittering of birds in the trees on either side. Strangely, apart from a few black jungle crows and an occasional flurry of sparrows, I saw very few of them. Now and again my attention was caught by a flash of exotic turquoise or yellow amongst the branches, but nothing so arresting as to warrant an ornithological essay. Perhaps a wider variety of birds would have been visible in another season, I don't know.

It soon grew wearisome trying to see over the tops of trees which regularly blocked my view, so I let my consciousness rest in itself, my gaze a narrowing beam upon the immediate few yards of trail which slowly unrolled before me. With a slightly inclined head I noticed things on the floor which I would otherwise have missed – or crushed into oblivion.

Once, beneath a thick trunk and shaded by its dangling roots, I spotted what looked like a fleet of miniature sails. Only when I came up close did I realise what they actually were – a dozen powder-blue butterflies at rest upon a rare patch of damp ground. I trod past the place gently, lest the vibration from my footfalls disturb them.

Before long I found myself nearing another high cliff rising from the water's edge. This time the trail, instead of detouring to one side or the other, continued straight across the rock face in the form of a gouged-out ledge. Protected by a knee-high wall on its outer edge, it was the sort of airy traverse which raised my spirits at the prospect of a little vertiginous excitement.

The path had been blasted from the 300-foot wide precipice in the early part of the twentieth century. A number of shot holes in the low roof bore testimony to both the obduracy of the rock and the skill of the anonymous workers who had created a safe passage that others might follow. Now and again, I paused to peer down through gaps in the parapet, calculating my chances should I stumble and plunge into the fast-flowing Ganga seventy feet below. The nature of the roof pushed me towards the outside of the ledge where it was easy to see how mules and sometimes people, disappeared from such relatively secure surroundings.

I soon reached the far side, where I paused to look back and take a couple of photographs. The harsh light had drained the scene of most of its colour and shadow, and I didn't anticipate good results. The Ganga, though, swept under the cliff in a broad, turquoise bend and was an impressive enough sight which would lodge in my memory for a long time to come.

As the afternoon crawled on, my thoughts turned occasionally upon the possibilities for the coming evening. A prosperous-looking village called Kandi, idyllically located on a terraced spur high above the river seemed a good place. A lane flanked by cultivated plots led to a cluster of whitewashed dwellings whose roofs shone like silver amongst the trees. 'Kandi'- the very name promised all that the passing traveller would need!

I halted at the turn-off and pulled out my water container to give me time to think. The plaintive bleat of a goat, the homely whiff of drifting wood-smoke, as well as the sound of lively voices nearby – they all sorely tempted me to lay down my pack for the day and put my feet up. I took a long slow pull from the wineskin, swirled the water about my parched mouth – and started walking again. It was too early. Only three o'clock in the afternoon; the pull of the ancient trail was too strong and I trudged on.

There is a beauty in farewells and I lodged the name of the village away in my memory for some future day when I might pass that way again. For a few minutes the sunlit roof tops and the curls of blue smoke remained in view over my left shoulder. But when I looked back from the first major bend, I saw only an unbroken canopy of trees. Soon the hamlet of Kandi was no more than a slight ache in my chest.

Even as I write about these events from a distance of two years, it is still a source of wonder to me how so many intense experiences could be compressed into so short a span of time. Less than three days! That was the way it was. One experience followed by another, like pearls on a necklace, each one succeeded by the next, and all contributing to the totality which I call my journey.

The day was already more than half over, but if I fondly imagined it would meander in leisurely fashion to its close, I would be disappointed.

I looked once more ahead and groaned inwardly at the sight of the trail ascending yet another sequence of stony diagonals. Fortunately the trees had none of the oppressive density and height of a true jungle, though the heat could hardly have been less pervasive. Thankfully, also, there were few flying insects to bother me. The thought of a cold drink, however, constantly nagged in my imagination and I found it difficult to avoid recalling the joyful dousing I had taken shortly before midday. Once more, in order to divert myself from such overpowering images, I called upon my 'folded tongue' yoga technique which I described earlier.

So intent was I on employing this 'cooling breath' in the recommended manner, that I all but failed to notice the motionless threat which confronted me!

A warning bell rang in my head and I stopped abruptly.

Not fifteen feet before me lay a large black snake blocking my progress. I say 'large' though I could only see a portion of if, perhaps four feet. The rest, including its head, was hidden within the darkness of a cavity beneath an embedded rock. The creature was about as thick as my upper forearm and perfectly motionless. Had the light not been good, I might have mistaken it for a branch.

I held my breath. Was the thing dead, asleep? Or what?

I took a cautious step forward.

At once a flicker of movement ran down its sinuous body. I could not tear my eyes away as a tremulous, elastic ripple followed. The snake began to reverse from the hole!

A foot…two feet…Three feet extruded from under the stone, slithered loosely upon itself – and stopped. Its head was still concealed, so as to its true length I could only guess.

A pause followed in which my first feelings of alarm were replaced by a riveting curiosity which could easily have been my undoing. The snake's menacing combination of grace, terror and raw elastic power had a hypnotic hold on my attention.

It started to move again, reversing once more under the stone. When it stopped, even less of its body protruded than previously. Perhaps it was simply looking for some quiet place in which to lie up for the afternoon? Finally, with one continuous flowing undulation, the snake poured itself into the hole and out of sight.

I thought it might be a cobra, though I couldn't be sure. When nothing happened for a full minute, I decided it was time to seize my opportunity while the going was good. I was aware that cobras have a fearsome reputation and didn't want to make any rash moves. The path was no more than six feet wide at that point, but because of the steep tree cover on either hand, a detour off the trail was out of the question. I would have to walk straight past the snake's den and hope for the best.

I clutched a handy stone just in case and stared hard at the hole. Snakes are ultra sensitive to vibration but I could discern no movement from within. Scarcely aware that I was holding my breath, I advanced a further three paces. In a moment I was beyond the stone – the danger was over. Or it ought to have been. Instead of dropping the stone and walking away, I then did a stupid thing.

Perhaps my judgement was clouded by tiredness but a mad impulse suddenly gripped me. Thinking I might discover whether the snake was a cobra – or, indeed, a king cobra – I lobbed the stone towards the hole. It fell short, rolled, and came to rest a few inches from the entrance.

There was no response.

Fortunately there were no other such stones easily to hand with which I might tempt further the 'cobra' to show its head. Finally, when the creature failed to respond I turned on my heels and lurched on towards Devaprayag. An image from my teens arose as I shuffled away – of a book illustration showing an angry king cobra chasing a galloping horseman. Maybe it was just as well that my aim hadn't been true.

The snake, or 'serpent', to give it its more emotive name, has a legendary if not awe-inspiring tradition in cultures throughout the world. In India, snake cults are symbolically connected with a variety of meanings which derive from certain of its physical characteristics, such as its energy, its undulating movement or its ability to shed its skin. Another reason why its sign expresses such diverse interpretation lies in the snake's varied habitat. Not only do they live in forests, deserts and heath-lands, snakes also lurk in lakes, swamps and wells, not to mention the oceans.

For Carl Jung, in *Man and His Symbols,* the snake, like other semi-aquatic creatures such as the lizard and sometimes the fish, is a symbol of transcendence arising from the depths of the unconscious (lake or ocean). He also points out how in the West the bird, too, has been used as an image of spiritual transcendence. In religious art from the Middle Ages onward, a composite representation of bird and serpent characteristics is not uncommon. The soaring flight of the bird (towards realisation, Bliss, Oneness with the Eternal and so on) is combined with the darker terrestrial image of destructive forces; darkness and light, the body and Soul – the higher is seen to be contained within the lower as the fully-grown tree is contained within the tiny seed. The winged dragon is a familiar example. Man, rising above earthly reality towards supernormal consciousness, is sometimes shown in winged flight. Because it can renew its skin, the snake also stands for resurrection and renewal, an ambivalence seen in the Gnostic wheel symbol (the Ouroborus) of the snake swallowing its own tail.

In the Chinese *Yin-Yang* symbol, the snake is clearly a profound representation of Change, the eternal rhythm of cyclical movement, with its alternation of light and darkness, the constant and the changing, action and inaction, the eternal in the now. The ancient symbol of a snake wound round a staff, or tree, has survived to the modern day as a sign of the medical profession (the Caduceus). The balanced entwinement of Mercury's caduceus, J.E. Cirlot suggests, is "indicative of an equilibrium of forces, of the counterbalancing of the cowed serpent (or sublimated power) by the untamed serpent, so representing good balanced by evil, health by sickness."

It thus foreshadows what Jung wisely noted – homeopathy's underlying philosophy, 'let like be cured by like'.

In the science of Yoga the coiled serpent is used to symbolically represent the concept of *Kundalini* or divine spiritual power. In Sanskrit, 'Kundal' means a coil and the serpent is thus commonly illustrated sitting at the base of the spine, or root chakra, in three and a half turns with the head looking up. Three turns signifies the three *gunas* or energies of nature which every individual possesses in varying degrees. For instance, a person deeply influenced by *Tamas* will display a torpid, lethargic and dull nature. A *Satvic* person, on the other hand, will have qualities of lightness, clarity and love and will not be swayed by external things. There will be much movement, activity and life energy flowing out in one who is dominated by *Rajas*. These three qualities prevent man from rising to higher states of consciousness. The raising of the *Kundalini* 'Mother' energy by practices such as Kriya Yoga, lifts the latent energy through all seven major chakras in the spinal column and completes the transformational process of uniting the individual with the Divine Principle. In most people, the 'serpent power' remains dormant for their entire lives; they remain 'asleep' to this ancient Tantric way to Self-Realisation.

It was around four o'clock in the afternoon when a natural lookout point on the trail gave me a welcome and uninterrupted view of the Ganga below. Beyond rose hills and ridges, the most distant melting seamlessly into the near-white sky. But something was not quite right!

I fished out my crumpled map to check.

The village of Vyasghat could not be far off, but why was the river flowing from the east when my map clearly showed a prominent swing the other way in close proximity to the village? There was not supposed to be a bridge across the Ganga at this spot. Yet there quite clearly was, unless I was hallucinating.

A host of irksome possibilities crowded my mind. Had I copied the map down wrongly...underestimated the distance? Maybe I had missed a turning? My pack dragged at my shoulders as I scanned the river in both directions – but I could see no sign of life on either bank. Already much of the wooded slope beneath me was beginning to fill with shadows. How many unnecessary extra miles had I unwittingly added to my journey?

Just then a movement on the bridge caught my attention: two bright splashes of colour were heading in my direction. Now I would soon know the worst! Ten minutes later two young women, both bearing loads, came up to me.

It was a particularly narrow place and I stood to one side to let them pass: "Vyasghat?" I added a few more halting words, my voice labouring.

The leading woman looked into my face from beneath her rolled-up load: her bangled arm wafted back towards the bridge: "Vyasghat."

I nodded my thanks and began to pick my way down to the crossing with renewed confidence.

Once across the bridge I turned left beneath an irregular wall of white cliffs running parallel with the river. As more villagers appeared I realised the village could not be far off. The initial recharge of energy which my encounter with the snake had given me had already begun to wear off and I was more than a little pleased when, half an hour later, Vyasghat's main street appeared directly ahead.

It was bigger than any of the other villages I had passed and I took an immediate liking to its cheerful early-evening bustle. Work in the fields was over and a mingling of voices, cooking aromas and neighbourly goodwill hung softly in the mellow air. Shadows were already beginning to lengthen across the street. By a little chai-shop I spotted an image of the monkey deity Hanuman seated within its vermilion-stained stone niche, where it bestowed a benign gaze upon the passing throng. A figure of undeniable magnetism to all who paid homage to it, the years and the respectful touch of many hands had blurred its features somewhat. However, after nearly ten hours on the move a rest seemed as good as a blessing and I flung myself down at the nearest table.

One of two men sharing a *hookah* wagged his head at my arrival and rose from a square-top table, served me and rejoined his companion. The only speech, a low bubbling sound, came from the pipe, as first one man, then the other inhaled deeply.

I sipped my scalding tea glad to be free of my pack for a while. A wooden railing enclosed the café's open front allowing me to observe the leisurely pace of mountain village life. Several scrawny chickens were pecking in the dirt by the restaurant steps, whilst through a window across the street I glimpsed a cosy domestic interior: a family seated on the floor, flickers of orange flame and a mother's voice mingling with that of a child.

It was good to sit idle for a while and feel the Western curse of over-urgency falling from my shoulders. Though a more critical observer of the passing scene might have concluded otherwise, my main impression was one of general happiness, if facial expressions are anything to go by. I am sure hill villagers in India can be as insular as anywhere else in the world, but there was a distinct air of relaxed goodwill, of a community in harmony with itself. Of course, on a wet monsoon morning with rain clouds down to roof top level, it might have struck me as a dejecting backwater, a place to leave with all haste. But as I sipped my tea, it occurred to me to wonder just how much we in the West have lost with our self-satisfied ideas of progress. It was clear that in the villages at least, the traditional ways still played a decisive role in maintaining a spirit of community. Would it last?

I pricked up my ears at an intrusive, unexpected sound.

A high-pitched, tinny voice rudely fractured the harmony, puncturing in a moment my Siva-tinted delusions, as an Indian pop-tune filled the air. The irony is that as Westerners in increasing numbers look to the East for inspiration, Asians themselves want what we profess to spurn.

In a while the feeling grew upon me to ask again of the English pair.

The café owner glanced down the street as if they had just left. His eyes brimmed:

"One very big Englishman, one very small. Yes, both men sleep my house last night; leave eight o'clock this morning. Big man walking very fast, small man often running. Very funny men - always laughing!"

I smiled at the image his description evoked. An hour of their jovial company wouldn't have been too obnoxious now that my own day was almost over.

The pipe bubbled languidly on as if time had no meaning.

After a pause I mentioned my encounter with the snake near Kandi. Both men listened intently, though I had my doubts that the second man understood what I was saying. The manager looked at me directly when I had finished and exclaimed:

"You very lucky man, sir – that snake king cobra! Every year many people die in India from cobra. Move very fast – most dangerous!"

I lapsed once more into silence, grateful that my aim with the stone had been awry. The *hookah* resumed its flatulent commentary.

Though I could still hear them, the chickens had drawn their own conclusions and had moved on. After the tea it was time for me to leave too, before my leg muscles cramped up. Heaving on my canvas monkey, I felt a leaden reluctance to move in both feet. "Relax…Stay here!" they seemed to shriek.

Nevertheless, almost before I knew it I reached the edge of the village and began to walk along the margins of a patchwork of green fields with boundaries of raised earth. If I had spotted a hotel in Vyasghat, I would have been glad to fork out for a night's lodgings, but I had seen no sign of one on the main street and now, against my better judgement, the trail had me once more in its tenacious grip. My map showed no more villages before Devaprayag, so just where I <u>would</u>

spend the night was an open question. The footpath, however, winding gently through thin tree cover very soon led me to a view so perfect that I stopped in my tracks. An exclamation of wonder broke from my dry throat.

It is difficult if not impossible to describe the sublime beauty of that silent, sunlit vista, a graceful bend on the Ganga's course, overhung by steepling mountains and forest drapery. Imagine a composition by Poussin or Claude Lorraine in which the various idealised elements of the painting majestically transmit a timeless, dreamlike vision – and you will begin to have some idea of the effect that place had upon me. In the foreground, with three or four stone steps leading to its arched entrance gleamed a small, white-painted shrine of perfect proportions and symmetry. My eye was pulled beyond it to the centre of the river, from which rose an imposing rock pinnacle, its knobbled top crowned by a delicate tree whose leaves twinkled like tiny firebirds in the faint breeze which stirred about the gorge at that point. Washed by the dashing water, the rock's dark base sparkled and shone as though it were an outcrop of precious ore. Above, rose fold upon fold of forest-clad mountainsides, each in perfect harmonic sequence until the final culminating ridge, an indented silhouette sharply etched against the evening sky. The entire scene about the river was dipped in an other-worldly golden light through which dainty butterflies flitted like wandering souls. The final piece in the composition, as though Nature imitated art, was the white-robed figure of a monk, reclining with his back to the entrance of the shrine. Lustrous black hair hung down to his shoulders. He was an imposing presence, so utterly without movement, he might have been a statue.

I held my step lest I intrude upon such saintly solitude. However, the monk, sensing someone near, half-turned and greeted me with a smile and a friendly wave. He was the first holy man I had encountered on my trek. I returned his greeting.

The distance between us was no more than thirty yards, but neither of us spoke. It was such a splendid, overpowering place that no talk was needed. With its cathedral-like mood of total calm it was an obvious place to call a halt for the night. It was one of those rare magical experiences when time past, time future, and the present, seemed to combine in one seamless vibration of energy. The place called to me to stop. The monk or sadhu – he would surely direct me to a suitable door, perhaps even in his own temple.

But a crazy stubbornness had taken over me... I hesitated. I was feeling drained mentally now, and despite the unforgettable scene, I couldn't summon the energy to carry me through the usual pantomime of asking for lodgings in a stranger's house.

I took a last lingering look around the garden-sanctuary. Part of me wanted to spend the rest of my days there; the other part pulled me away. I glanced finally at the bright curve of water whose flying motion complemented the silence of hills which rose in veils of green above it. Next moment, the Arcadian scene, forever lodged in my memory, was behind me as I turned my back and set off once more.

I strode on, my mind now a slow-motion jumble of shrinking options as to how I might spend the night. Some five hours had now passed since my rest in the dry stream bed; my snacks were all-but finished and I was beginning to feel hungry again.

At the first hint of an uphill gradient my pace slowed as the muscles of my lower legs began to complain. Once, I stopped to rub a pressure point below each knee in order to relieve the dull ache of my limbs. Japanese soldiers used the technique during the 2nd World War on long marches, but with one difference: instead of applying pressure with a finger or thumb, they were reputed to use a lighted cigarette end.

It helped a little.

Tired though I was becoming, it was as if some force was pushing me on and on towards Devaprayag. It was out of the question that I could reach the town before nightfall, or even midnight. Still I kept on. The sheer repetitiveness of the act of walking, of planting one foot mantra-like after the other, again and again, induced in me a temporary trance-like state, in which my awareness shifted between identification with my physical discomfort and a more subtle state in which I seemed to glide effortlessly over the ground. Unfortunately, the latter, more enviable condition would last only a short while, before being overtaken by the former. It was my own fault, really, for setting such a punishing pace in which distance and nearness blurred into one; inner and outer worlds slid into one another. If only it had lasted! I didn't want to admit the fact to myself, but my ego was hooked on the absurd notion that I might catch up with Laurel and Hardy round the next bend or find them chuckling over a beer in some rest-house along the way. After only twenty minutes walking I spotted a stone building directly ahead amongst the trees.

I came up to it shortly and saw it to be the front of a crumbling, temple-like structure, some ten or twelve feet in height. In its centre was a solid-seeming wooden door. Bushes and trailing tangles of branches encroached densely at either side of the wall, hiding the rest of the building from view. Whether it was the effect of a growing physical fatigue I cannot say, but I had the curious feeling that nothing lay behind the door. It was a kind of folly, I thought.

Whatever the cause or reason I pushed this thought to one side and began to make towards it – a benevolent hand had obviously guided me to this ancient porch. Within, I would surely find all that I desired at that moment: water, food and a bug-free bed – in that order.

But before I could take more than a couple of steps, an unexpected thing happened.

A tremendous commotion broke out behind the closed door.

Next second it flew open – and two hoary old *babas*, naked save for a ragged loin-cloth apiece, rushed out gesticulating and shouting angrily at the tops of their voices. A third man, much younger and dressed in a clean *lungi* and *kurta*, or long shirt, appeared behind them.

For a moment, as the emaciated pair bore down on me, I thought I was the focus of their wrath – but there was no time to think as they were almost upon me. Only when one of the grey-bearded beanpoles stumbled, did I realise that their shoutings were directed not at me- but at each other. Both men looked blinded by the passion of whatever it was they were arguing about. The younger man, meanwhile, had seated himself on a low wall and took no part in the furious altercation. There was no aggression in <u>his</u> attitude – on the contrary, his presence was reassuring in the circumstances. Glancing vapidly at me, he noted my appearance and, in a strange, sing-song voice, observed: "No, I am sorry – it is too late go Devaprayag…it is too late, Devaprayag."

His eyes were red-rimmed like those of the other two, making me wonder if all three of them were not high on *ganja.*

Whether the man's opinion was an invitation for me to stay the night with him and his friends – I could not be sure. Nevertheless, the very thought of spending the long hours of darkness in the company of two such madmen jerked me back into life: I would surely sleep more peacefully out in the open than cooped-up with that dubious pair.

As I trudged on, they formed an indelible impression in my memory – two spectral bodies dancing amongst the trees like a pair of drunken marionettes. Their rantings followed me for a while, but at last I was out of earshot.

The incident served only to sharpen the need for me to answer the question: where would I spend the night? Should I attempt to sleep outside again? Perhaps I would chance upon another rest-house if I kept going? As a last resort I could continue to walk on slowly through the night, trusting to my instinct for finding the way. But I wasn't all that enthusiastic about such a solution; my dry throat and a leaden feeling in my feet told me I would do well to rest – but where?

Numerous possibilities, some more bizarre than others, tumbled through my mind as evening twilight closed softly around me. To the west, the marigold-yellow sky had by now faded to the palest lemon, whilst on the opposite side of the Ganga, the mountainside was already a velvet wall of deepening shadows. Night falls rapidly in these parts; another hour and the first stars would begin to emerge. Fortunately, an answer was not long in coming – or so I thought.

When I came to a break in the trees with a gentle green slope leading down to the river – I dropped my pack with an exclamation of excitement: this was it – the perfect place!

From the trail, a narrow indentation snaked through the short, animal-cropped meadow to the water's edge. I would take a swim, wash, but first things first –

I gathered some twigs and sticks from the margins of the footpath and soon had a brew of tea prepared. I followed this with the remains of a packet of mushroom soup and my last breadcrust. As this two-course banquet of frugality was cooking I remembered with a twinge of regret the energy-food, '*prana*-balls', which I had rashly decided to leave behind in my room.

I had spent some time preparing a dozen of these 'solar powerpacks' at home. Their preparation was simple enough: a variety of dried fruits chopped fine, mixed with cashews, sunflower and sesame seeds, desiccated coconut and carob powder. The binding agent was a mixture of honey and raw oats. Having no honey at the time, I added a little water, instead. One *prana*-ball, according to the American wilderness bible, weighing five or six ounces, would supply one day's trekking energy. Whilst packing in my ashram room I had decided at the last minute to omit them in order to save weight. I also reasoned I would be much more in need of them when I journeyed to the higher, lonelier regions, where food would be in even shorter supply. Besides, 'mere foothills' would surely not extend me too much? My miscalculation now meant that tea, thin soup and the last of the 'baby' biscuits would have to serve me until Devaprayag.

The minimal food served its purpose in refreshing me however, and I began to feel more positive about sleeping in the open. Only when I started to tread out the fire's embers did I notice anything unusual; my 'picnic' spot was not as perfect as it had seemed at first glance.

A rim of greyish sand flecked with mica at the water's edge, the stand of tall, dead-looking trees and the slow swirl of an eddy would have normally been a safe place to swim. In my freshwater fishing days, an eddy was considered a productive place to cast your line, especially in winter when a river generally runs fast and deep. Fish would often take up position in the eddy, just out of reach of the main drag of the current.

Now a different kind of fish occupied this sluggishly turning loop of soupy water; three greyish-white shapes swung slowly in the slack water. Round and round they drifted with never a flicker of life, a slow, gruesome procession.

On the swollen, glistening midriff of one of the corpses a solitary crow stood motionless, its head cocked sharply like a blackbird listening for worms. Above, in the topmost branches of the skeletal trees, a gallery of vultures waited in patient, sinister vigil, their hooded eyes hooked upon the waterlogged creatures a short swoop below.

I wasted no time in further conjecture, but stuffed my belongings back in my rucksack and hurried on, glad to be leaving the macabre scene behind. I began to think I was destined to walk all night.

The forest through which the trail cut its way, was now assuming a much lonelier aspect as darkness closed around me; trees, with branches twisting this way and that, tangled roots and boulders, all began to strike intimidating forms. Unfamiliar noises started out of the shadows, and it was all-too easy, if I allowed it, for my mind to dwell upon the cobra incident. This, the largest poisonous snake in the world, sometimes exceeds twelve feet in length and is a creature of nocturnal habits. What other animals might roam the trail during the long hours of darkness, I didn't care to think about.

An American Theravadan monk once told me of facing his own fear in similar circumstances deep in the forests of Thailand. Part of his training included spending several nights sitting alone under a tree with no protection except his own mind and a small candle. He heard a tiger roaring in the distance and sometimes was very scared. Only when he lit the candle and sat within its circle of light did his fears subside to a manageable degree. "Beyond the light was danger; within was security", he told me.

I was carrying a small torch but preferred to save its batteries for times of real necessity.

I plodded on for a further hour through the darkness without seeing a soul or friendly light. Although stars wheeled overhead in a blue-black sky, at any one time I could see no more than twenty yards of the faintly glimmering trail ahead of me.

High amongst trees on the other side of the Ganga, however, I spotted the flickering glow of a small fire – fire amongst the mountains meant food, talk and security from roaming animals. But there was no bridge in the vicinity so I walked on ruefully, alone with my own shut-in thoughts, but always alert for the strange sounds about me. I began to experience mild hallucinations. Whenever I blinked, the darkness responded immediately with a hundred pinprick flashes of light. An agnostic epitaph popped up on my mind's screen: "Destination unknown". I laughed silently, uneasily.

A little while later when my unease was at its height, there occurred one of those 'coincidences' to which I referred to earlier.

On the spur of the moment I began to chant silently the sacred Hindu mantra *OM* (pron. AUM). I had known of this mantric sound for some time, but had been reminded of its power when Dharmananda made reference to it during a talk:

156

"We are all accompanied through life by a guardian spirit. Whenever you are low, whenever you are down, chant *OM*. Help will come."

In the past I had not been conscious just how far one's inner state could be fine-tuned, sometimes with dramatic, long-lasting results, by repeating such syllables. It is probably just as well because I would have used it in a mechanical way with no understanding of its more profound levels of meaning. Krishnamurti dismisses the value of mantra chanting as a sham, claiming their effect is to produce no more than an ox-like calm, which evaporates almost as soon as the practice is stopped. All I can say, is that one should try such things for oneself and observe the results.

For Hindus, particularly yogis, the mantra *OM* is said to be the sound symbol of God and is thus considered to be the most perfect sound (as God is considered to be perfect). The sound derives from the understanding that for anything to come into being, Existence, there has to be a creative vibration. The potent sound *OM* is regarded as the nearest we can come to this cosmic vibratory power. The belief – though it is more than the bare product of speculative thought – is not so far-fetched as it seems.

There are similar parallels to be found in other religions as well.

In Tibetan Buddhism, for example, the mantra, *OM MANI PADME HUM*, is chanted wherever Tibetans are to be found. For them the act of repeating the sacred syllables coupled to the turning of the prayer wheel is, however, not an act or prayer as a Christian would intend it. Neither is it an act of crude self-hypnosis. The Tibetan is not trying to invoke some Power outside himself for help or personal gain, but 'gathering to a point' his own inner forces. When this is done effectively, the sounds begin to acquire a deep materialising power. One becomes conscious of one's place in the universal rotation, the Cosmic Law; and what needs to happen in one's life, begins to happen.

The Word of the Bible, *AMEN*, means in Hebrew, 'sure' or 'faithful'. It is the sacred word of the Egyptians, Greeks, Romans, Jews and Christians.

"These things saith the Amen, the faithful and true witness, the beginning of the creation of God (Revelations 3:14)".

Moslems have the symbol-word, *Amin*.

The danger of lifting such 'power-mantras' from their ancient contexts is that they lose their spiritual substance; they become no more than a tool in the secular hands of the Western social adjustment industry, a quick-fix mental aspirin to psychological 'growth'.

Whatever the explanation for the forces which are set in motion through the chanting of a mantra, unless it arises out of a heartfelt conviction, there is likely to be little or no profit from its mechanical repetition.

To cut an interesting diversion short, an unexpected consequence followed quickly on the heels of my starting to chant the mantra. My gaze, as I repeated the syllable to myself, was so intent upon the few yards of path which were visible to me, that I failed to notice another break in the trees bordering a strip of rough pasture by the Ganga.

A man's sudden shout jerked me from my solitary musings upon the world's ills.

"Come! Morning time, Devaprayag!" Twenty yards from the footpath, a man wearing a crimson turban and sitting against a ten-foot log, beckoned to me again.

Was it so obvious to everyone where I was heading? But, needing no further encouragement, I breathed a prayer of thanks and stumbled towards the fire before which the man was sitting. More than twelve hours walking was a long stint and the shepherd's summons was like a call from heaven.

At my approach, two more men arose from their watching stations amongst their flock to greet me with hospitable smiles and a few words in the vernacular which I didn't catch. Both were dressed in the customary hill-shepherd attire of the Tehri-Garhwal region which included a loose shirt, tightly fitting waistcoat or jacket and a round, flat-topped hat. In the firelight I could see by the state of their footwear that they were not well off, at least by Western standards, but I immediately felt at home with their cheerful demeanour and knew I need travel no further that night. The man in the red turban indicated with a wide sweep of his arm that I should sit down with him. With his well-defined features, shrewd eyes and a short beard flecked with grey, I took him to be the leader.

The clearing was filled with some two hundred goats, all clearly in the process of settling down for the long hours of darkness ahead. Several of the foremost animals eyed the stranger warily as he sat down. However, I obviously presented no threat as shortly, their eyelids drooped again, a few plaintive bleatings died away and a calm once more settled over the dreaming flock of bearded ruminants.

The leader's eyes gleamed as he pointed to the sombre shadows from which I had just emerged: "*Janvar!*" He made a menacing claw of his hand and was clearly warning me of the dangers which lay beyond the protective circle of the campfire's light.

"Tiger?" I nodded, mirroring his mime.

"Shér," came the short reply; the other two wagged their heads in sober fashion.

After a pause, I made an act of wiping my salt-dried brow and pointed to my well-travelled boots: "Devaprayag, morning time – *chalna.*"

The shepherds chuckled in unison. One of them disappeared, returning a few moments later with a tin cup of warm goat's milk. I soon realised just how dehydrated my body had become and drained the cup in two gulps. The shepherds recognised a thirsty traveller when they saw one, and my cup was no sooner empty than it was whisked away to be refilled, I took the second cup more slowly, pausing now and then with a mainly comical but unsuccessful attempt to converse with my new-found friends.

Indicating the large pot gently simmering on the fire, the leader made eating motions with his hand.

"*Khana?*" I observed, this being one of my short list of 'essential' words of Hindi. I felt hungry and ready for bed in equal portions, even though the time was only just after eight. He patted my rucksack which I was using as an arm-rest: "Country?"

I pointed vaguely west, in the supposed direction of my rented home some six thousand miles away: "England," I responded. "London."

"Ah, *Angrezi*." The man wagged his head in that brief movement I had already become accustomed to in India: part nod, part shake of the head – nothing is quite what it seems to the Westerner trying to make sense of his experiences in such a contrary, diverse and colourful country.

I would have been glad to find out something of those men's lives as mountain dwellers – if only I had known more of their language. However, my ignorance of Hindi didn't seem to bother them for on the contrary, they seemed to appreciate the efforts I did make. I showed them one or two items from my pack, such as my French clasp knife, which I thought they would find of interest. I was grateful that they expected little enough of me that evening. In truth, they seemed happy with minimum conversation between themselves; otherwise they sat in silent absorption, blending with the terrain – sitting, yet instantly alert to the occasional squawk from the resting herd.

Almost without my noticing, the moon had risen over the hills behind the camp, throwing a silvery light across the silent Ganga and the wooded steepnesses above the opposite bank; the bank-side on which we drowsed was bathed similarly, in the same shining light. Whatever had brought me to this perfect spot, it was easy to believe that the invisible spirits of the ancient *rishis* dwelled there. At any moment it seemed, Siva, Krishna or one of the other countless deities would step softly into that enchanting, sylvan scene.

Nearly two more hours slipped by before the shepherd indicated to me that the food was at last ready. He called into the darkness, whereupon two more men silently appeared and sat down with us in an irregular circle.

I was treated like the guest of honour and served before anyone else: a battered plate carrying a wallop of white rice the size of a family Christmas pudding was handed to me with a display of touching courtesy which belied the shepherds' rough exteriors. *Dal*, a thickened lentil soup, a *subzee* of potatoes and a wad of warm chapatis soon followed. It looked and smelled tasty, even if it did bear a striking resemblance to the previous evening's fare…and the evening before. The shepherds would be eating it for the rest of their lives.

The family cutlery was not on offer, so, churning the ingredients with my fingers I followed the usual custom of employing the right hand for eating purposes.

Only when I pawed an unwieldy wodge of the mixture into my mouth did I discover that the meaning of 'hot' is a very relative term when applied to Indian cuisine. Within seconds my mouth and throat were on fire. I gulped and made a grab for one of the metal water cups. The contents, however, gave me only a temporary respite from the searing spices.

To refuse to eat more, as if I was in my own home or a Western restaurant – that would only cause offence. Tears sprang from my eyes and coursed down my cheeks; sweat beaded my forehead where none had been for several hours, yet I forced myself to eat a good half-plate of the lethal fire-mixture as well as all of the chapatis. At that point my stomach felt incapable of holding another gram and I took the water with measured sips. Fortunately, rescue was at hand.

A younger shepherd with a sparse black beard, sensing that my streaming face was not entirely the result of unalloyed joy, gently took the plate from my grasp. His eyes wore an understanding twinkle when he replaced it with a small bowl of plain, cold rice, milk and sugar. This episode caused much amusement all round and fortunately I was able to appreciate the funny side as well.

It was now eleven o'clock and I had not budged from my backrest for three hours; my eyelids were beginning to droop heavily with the desire for sleep. I glanced round the circle, but no-one except me looked to be on the point of retiring: I would have to sit it out and await developments. The men were all hardy types and I started to wonder if I was to be part of an all-night vigil. If that was the case I would have to use my sleeping bag against the night's chill which was beginning to feel its way through my cotton garments.

A little later, when some of the more distant animals started to bawl, one of the men sprang to his feet in animated fashion and stared intently into the darkness to my right where some animal might well be stalking his flock. But on this occasion it was a false alarm and shortly he sat down again, satisfied that all was quiet.

Then began a short, embarrassing episode for me – one I could so easily have avoided had I not spent all my small denomination notes in the tea-shops along the way.

Despite my increasing drowsiness, I noticed that two of the shepherds were showing considerable interest in my footwear. Although nothing special by Western standards, I supposed that my scuffed trekking boots must have marked me out as a rich foreigner in the minds of these hardy souls, with their ragged jackets and worn, army-type plimsolls. It was useless for me to explain that I lived close to the breadline at home on an income of half the national average. The fact that I could afford to visit their country – by plane – automatically established a division between us in their eyes.

The two exchanged comments between themselves and with the leader who was still seated on my right. Then the one with the black beard, noticing my awareness of his gaze, thrust one of his own feet into view. Even by the flickering light of the now dying fire, I could see plainly that the young man's bare foot protruded where the rubber sole had parted from its cheap canvas upper.

He looked at me and opened one hand in a half-humorous gesture of resignation.

I responded with a few mumbled words of commiseration. But the situation all too clearly called for a more concrete response on my part. The shepherd in the red turban pursed his lips ruefully, though fortunately he made no formal request of his flagging guest.

If only I had some low denomination notes – I would gladly have parted with twenty rupees, or whatever it took to buy the shepherd a new pair of plimsolls. All that remained in my money belt, however, were two, Rs100 bills; I was sure to need them if I took a hotel at Devaprayag and a bus ride back to Rishikesh.

It was an embarrassing situation which I could have done well without, particularly as my brain was too spent to think of some alternative way of showing my appreciation of their generosity.

I thought I detected a *frisson* of expectancy in the air, but that might well have been merely a projection of my own guilt feelings. I am ashamed to say I did not rise with due magnanimity to the occasion by parting, as one hears of others doing, with some prized possession which the recipient would treasure for the rest of his mortal days. No, I simply offered round my remaining *beedis* and feigned ignorance.

There was an awkward pause or two in the murmuring between the two shepherds, before the young man slowly withdrew his foot: the matter, I sensed, was being discreetly dropped.

What seemed like another eternity dragged by before three of the shepherds rose to their feet. One of them mimicking sleep by touching the tips of his fingers together, indicated that I should follow.

I breathed a sigh of relief.

Together we picked our way through the flock, traversing the bank until we came up to a narrow suspension bridge some hundred and twenty feet long. The shepherd who had shown me his shoe gestured towards a square stone hut shining with supernatural luminosity on the far side: "Sleep," he informed me gently.

I nodded and was thankful.

The interior of this temple of dreams was just large enough to allow four of us to sleep side by side on the floor. Apart from footwear, I was the only one to remove any clothing, a mistake as it turned out. The others merely stretched out full-length and threw a single large blanket over themselves. In contrast, I unrolled my down-filled cocoon of a sleeping bag which drew appraising glances in the low light of the lantern. But I was past caring that I might wake up in the morning without it - and as soon as my head touched my improvised pillow, I fell asleep.

The remainder of the night passed in a dreamless trice as I slept the sleep of the dead.

I awoke to find a solitary sunbeam shining through a chink in the wooden door and the other three already up and about. Naturalists' books of facts will inform you that India is home to thirty thousand species of insect; the blotches on my exposed upper arms bore testimony to a nocturnal acquaintance with at least one of them: the ubiquitous flea.

I lay on my back for a few minutes more, trying not to scratch the ugly red weals and listening to the cacophony of bleating outside. Eventually, I could suppress my curiosity no longer, struggled out of my sleeping bag with a yawn, packed my things and wandered outside.

A very different scene met my gaze compared with the evening before. Now, instead of stillness all was movement and noise. On the far side of the Ganga, goats were streaming towards the footbridge from every direction; the undergrowth seethed and shook with their passage, their cries filling the early morning air of the gorge and mingling with the shrill whistles and sharp instructions from the shepherds. Already a procession was straggling across the bridge urged on by a shepherd at the far end.

I took the opportunity to wash at a spring close by the hut. One of the men had remained behind and was cooking breakfast over a low fire in the lee of the wall.

In a little while we were joined by three others and I ate a brief but welcome, breakfast of tea with chapatis reheated from the previous evening. By the time I had finished, most of the goats had crossed the bridge and were dispersing along the innumerable trails on our bank. It was time for me to say goodbye. I shook hands with all of them in turn, each wishing the other well in his own language. Then I was heading for the river and the Devaprayag trail.

Hardly had I regained it, when I heard a shout from high up. I stopped and looked back. The hut by now was an island set in a sea of swirling goats and at first I could see no-one. Only when I searched the hillside high above did I spot the turban and the tiny arm waving in farewell. Who knows, perhaps one day I will pass that way again and be in a better position to return the kindnesses I had been given so freely?

For a few minutes a tinge of sadness clouded my mood as I walked on through the grey-green hill-cover. But before long my spirits rose again and I heard only the regular rhythm of my boots on the path and an intermittent creaking from my rucksack shoulder straps. Devaprayag could not be more than a few hours away by now, so I no longer felt driven by the blinkered energy of the evening before, to walk fast. I hiked on, pausing only for an occasional drink from my water container. Once I stopped, peering steadily across to the opposite bank where I thought I spotted some creature or other standing amongst the boulders littering the shoreline. Only when the 'something' hadn't moved after a full minute, did I decide it might have been an accidental shape conjured up by Nature.

Easy to dismiss as no more than a result of personal whimsy. Yet the mystical poet, Antonin Artaud, who in 1936, travelled through Mexico in the mysterious land of the peyote-eating Tarahumara, wrote:

"*The land of the Tarahumara is full of signs, forms and natural effigies which in no way seem the result of chance - as if the gods themselves, whom one feels everywhere here, had chosen to express their powers by means of these strange signatures*".

As the miles shuffled behind me and the sun's rays inclined more and more to the vertical, the landscape gradually assumed a much starker aspect. Where the lower slopes had earlier been covered by a mantle of trees, more and more bare expanses of rock now thrust through the steeply tilted hillsides.

After walking for three hours I came within sight of a 100-foot cliff which seemed to bar the way to any further progress along the river bank. But I needn't have concerned myself for, as is often the case, a cliff can seem much steeper than it is when viewed *en face*. At close acquaintance I found it not so steep as I had imagined and to be broken up in a mounting series of crumbling blocks divided by runnels of fine scree. Five minutes of dusty scrambling took me to the top, which proved to form the edge of a sweeping dirt road. I realised that my walk must now be nearly over.

So it proved. An hour later as the road began to descend in a sharp bend, I suddenly saw the town of Devaprayag directly ahead. The two rivers Bhagirathi and Alaknanda both converge at the town to form the true Ganga. My vantage point was an ideal place to call a short halt.

It has to be said that after all that had gone before, arrival was something of an anti-climax. At the moment the journey was over I felt no great emotion. No thunderbolt from Siva fell from the heavens to mark the occasion - He was obviously absent from the 'prayag' or 'sacred confluence' attending to more important matters. I stood beneath the paltry shade of a twisted tree for a few minutes, crunching a celebration broken biscuit and absorbing the abrupt confrontation with so many buildings. Devaprayag appeared to be asleep.

In a short while, I took a final gulp of water, before trudging on to investigate further.

Soon the road brought me to the head of a lane down which I could see a narrow metal footbridge spanning the Alaknanda, eighty feet below. The town is built on a serenely picturesque spot. Over aeons of time, the two rivers have carved three large 'knuckles' of rock round the *sangam* or holy place. Double and triple-storey dwellings rise up the hillsides in serried rows, divided and linked by steep lanes and unexpected flights of stone steps. Many of the houses have wooden balconies overlooking the rushing waters below. Other bridges interconnect with the rest of this important township, from which two roads diverge, one to the pilgrim place of Yamunotri, along the banks of the Bhagirathi, the other following the Alaknanda to Kedarnath, an equally holy destination for countless Hindus over the centuries. Gangotri, the last village en route to the source of the Ganga, is also approached along the dramatic Bhagirathi gorge.

The narrow lanes of the town are unpolluted by motor vehicles and all goods, including enormous sacks of rice and other grains have to be ferried from one of the roads, by sturdy porters with legs of knotted sinew. It felt like Sunday as I wandered along. Few people were about and it was twenty minutes before I finally stumbled on a *dhaba* which was open for business. After a supine hour at a table within its cramped confines, I set off to pay my respects at the main *ghat*.

Another tilted lane full of shadowy warehouses, shops and the open doorways of people's homes soon carried me to a broad fan of steps above the age-old bathing place. I use the word 'bathing' advisedly for the force of the torrent was such that to leave the security of the lowest step would mean being instantly plucked away by the tumultuous current. Apart from a snowy-maned sadhu washing his clothes, the steps and promenade which formed the *ghat* were deserted. It was a very different scene to those I had witnessed at Haridwar and Rishikesh. I decided to follow the other man's example and wash some of my own clothing at the water's edge; my insect bites and hot feet were also badly in need of a cooling dip.

The sadhu ignored me as I unslung my rucksack and kicked off my footwear with a quiet feeling of journey accomplished.

Next moment I was knee deep in the Bhagirathi, 'the sacred thread', a few yards short of its actual meeting with the Alaknanda, where the force of the river was at its fiercest. The freezing glacial water was like an icy kiss on my legs, but a treat all the same, which I had been anticipating for several hours. Close by, a ten-foot metal chain as thick as my arm was stretched out taut by the ceaseless pull of the relentless current. For the more devout pilgrim or venturesome thrill-seeker, great spiritual merit would accrue from hanging from this weed-slicked chain at the very confluence of the two famous rivers. I took a short look and decided discretion to be the best option.

The confluence is a place rich in symbolic myths. According to the ancient legend, Northern India was once the kingdom of King Sagar, father to sixty thousand sons. Perhaps having nothing better to do with their time, the sixty thousand sons were one day disrespectful to a venerated holy man called Kapil Muni. He was so infuriated by their impertinence, that with a single searing look he reduced all the sons to ashes. The king in his sadness turned to the gods for help and they told him his sons could only be brought back to life if their sinful remains were purified by the water of the Goddess, Ganga, who in those days was a river flowing through Heaven. Acting on the king's behalf, another holy man, Bhagirath, pleaded with, and finally persuaded, Ganga to descend to earth. At that point Siva intervened, allowing Ganga, the holy river, to tumble gently through his hair to Gaumukh high in the Himalayas, and ever since venerated as the source of India's greatest river. From there, Ganga then cut deep gorges through the mountain barrier emerging at Haridwar, before eventually reaching the ocean some 1500 miles away in the Bay of Bengal. There she washed the sixty thousand sons' ashes and returned them to life. Since that time, a festival has been held every January on Sagar Island at the mouth of the Hooghly River to commemorate the occasion.

There was nothing I would have enjoyed more than a complete bath in 'Siva's hair', since the sun was now directly overhead and almost too hot to withstand on the unshaded *ghat*. But I could do without the kind of challenge presented by the streaming chain. I remembered for a moment the man who vanished at Laxman Jhula. Had he been struck by a reckless impulse, and too late, tried to change his mind? For a few seconds I understood my own frail hold on life, as the ice-chilled water trickled down my shins to evaporate on the burning stone. Who is it that dies, anyway? Who is it who changes?

As I ate the last of my apples my mind slipped into a more worldly gear. Tehri, the location of the much-debated superdam was only fifty miles distant – what effect would its completion have upon the spectacular volume of water rushing past my feet?

In my teens I saw a picture in *National Geographic* of Colorado's giant Hoover

Dam, the first of such monoliths and the prototype for similar such schemes around the world. Even then, I felt a strong sense of rebellion against the smugness of its boastful caption:

'Taming A Once Mighty River'.

What are the alternatives if indeed there are any?

At Adgaon, a village in the state of Maharashtra, 240 kilometres south of the Sardar Sarovar superdam on the Narmada River, a method of 'rain water harvesting' is being tried. A number of small earth dams have been constructed across the streams of the district using local materials. The villagers have also built earth walls, or *bunds*, around their fields in order to control rainwater run-off. The rain is thus contained largely where it falls and can be stored during the months of drought. It is also cheap to build and maintain. Farmers can also later recover the fertile salts which have washed down behind the dams and use them as fertiliser on their fields. Fish have begun to appear in the streams, whilst groves of acacia, eucalyptus and banana trees have been planted with every prospect of making a flourishing contribution to the economy.

Apart from the little matter of building superdams in earthquake-prone regions, one of the main fears of opponents is that the small local communities will lose out. The huge cost of constructing canals and aquaducts to transport the water to areas far away, may outweigh the benefits, according to critics, especially as such canals already in use leak badly. Old ways of life are being lost forever, though some tribal peoples are making a stand, declaring they would sooner drown in their own homes than agree to resettlement elsewhere.

It might have been the time of day, but the up and down lanes did seem uncannily quiet compared with the lively Vyasghat of the previous afternoon. It was hard to believe that before the road was built some seventy years ago, fifty thousand pilgrims a year would pass through Devaprayag on their arduous treks up the 'sacred threads' to the holy shrines of the high Himalayas. Today, the road has brought with it the tourist along with the true pilgrim. For most motorised travellers, the junction is now no more than a rest-stop along the way: a meal, a leg-stretch – then onward, leaving the perched little township to slumber on undisturbed, dreaming of its past glories. For the first time that day, I found myself musing upon the two travellers from my own country. Had they arrived? Perhaps they were already on their way back to Rishikesh? Somehow I couldn't see them carrying on beyond the joining of the waters.

I decided that now would be a good time to rest-up for a while by the Alaknanda and accordingly began to retrace my steps towards the first suspension bridge which had given me access to the main part of the town. It was two p.m., still too early to go in search of a room.

I recrossed the bridge and a steep, scrambling descent soon brought me to a stretch of sandy beach, empty, apart from a very un-heavenly scattering of dried dog turds. I made my way towards a prominent rock, veering off to a smaller one at the last moment; having spotted a stretched-out human leg I didn't want to intrude upon somebody else's privacy. The alternative wasn't ideal, but after a moment or two of indecision, I flung myself down in its meagre shadow.

Despite the sheltering wall and the proximity of the river, there was little respite from the broiling sun. I lingered there for two hours as it crawled slowly across the cloudless sky; but not for a moment did a cooling breeze fan the dry air. I hadn't showered in three days and felt the urge for a room of my own before I could properly relax. I knew there was a Tourist Lodge on the edge of Devaprayag and forced myself into action.

Being an inventive sort, I took a steep, direct line up the slope to the footbridge. It was supposed to be a short cut, but I soon found myself toiling through bushes and trailing thorns which dragged at my skin and clothing. A conical tip of household refuse greeted me as I emerged from the thorns, and I mounted this with a tremendous clatter of rusting metal and other detritus of domestic life. I recrossed the bridge after a deep-breathing halt and a pull from my water container.

Shortly I spotted the sign I had been hoping to see.

As I followed the arrow, a short muscular porter trotted past me with a grunted warning, his face half-hidden beneath his load of grain. The enormous sack must have weighed close on two hundred pounds – only a little less than my own body weight and rucksack combined!

Plodding up the slanting lane, the rounded tops of the neighbouring hills began to appear, their flanks patterned by rhythmical tiers of contoured terraces which must have taken many generations to complete; the slopes seemed almost to be in movement such was the effect of sinuous repetitive walling upon the eye. An innocent-enough district today but it was a very different story around 1918, as I have already mentioned in an earlier chapter. Preying on the local populace between Rudraprayag and Devaprayag, the marauding, man-eating leopard killed over 120 people not to mention the many domestic animals, before Jim Corbett finally caught up with it late one night.

Much of the township dates from medieval times and I had anticipated the Tourist Bungalow to look something like its brown and white painted neighbours. Instead, it stood out like a sore thumb, a functional, no-nonsense glass and brick structure, an incongruous foreigner not even masquerading as an original. With its façade of spotless brickwork and vacant windows, it resembled a piece of real estate waiting for a buyer. It obviously wasn't built to attract penniless sadhus and

I hoped I had enough money and wouldn't be turned away for looking too scruffy. I was mildly surprised to find it open.

Within, I found myself entering a cool, spacious interior with a bare noticeboard and an unoccupied reception desk. I rang the polished bell. Nobody came. The building was as silent as the inside of a pharaoh's tomb, causing me to wonder if it really was open.

I rang again.

Eventually, after an interminable wait, I heard scuffling movements in the background before a young man materialised from the shadows. I think I must have awoken him from his afternoon slumbers, but he treated me with enough regulation politeness to make me think the place wasn't so bad after all.

Just as he was handing me my room key, however, a totally unexpected thing happened.

From the far end of a sterile, college-like corridor, a staccato burst of maniacal laughter split the silence. It made me think of asylums, locked doors and padded cells. The disturbance was over as abruptly as it began, and a monastic quiet descended one more upon the lodge.

"Very funny people," was all the manager said, giving me a bemused glance.

My room was large enough, spartanly furnished and without a speck of dust upon its pristine stone floor. A poster or two would have cheered the place up, but I was too much in need of a good wash and a proper rest, to concern myself with such civilized niceties. I made a fist and struck the wall a sharp blow adjacent to the switch panel. The response was instantaneous: a short <u>click</u> and the fan blades began to revolve with increasing speed. I turned it off, stripped down to my money belt and slung my towel about my waist. Down the corridor I found the shower working at no more than half-pressure, but it gave me a much needed lift as I soaked the aches from my shoulders and the rest of my body. Afterwards I lay down under the fan on the ascetically hard bed intending a few hours' rest.

At that very moment, however, all such fantasies vanished as my rest was rudely shattered by another ear-splitting roar of satirical laughter. It came from near the shower room and this time was accompanied by a string of muffled repartee in my own language. I poked my head from the door to hear more laughter and the sound of objects being thrown around. There was only one thing for it.

Retracing my wet footprints I halted in front of the penultimate door before the washroom. The vigorous commotion continued from within, a roof-raising mixture of fragmented sentences and answering hoots of mirth.

When I knocked sharply the shrieks ceased as instantly as turning off a tap.

A second or two passed before a man's voice, sobriety wrestling with hysteria, growled with assumed indifference, "Yeah!?"

Light exploded into the corridor as I swung the door open and stared into the room. There before me, sprawled across their separate rumpled beds, surrounded by a shambles of half-devoured food, miscellaneous personal belongings and trekking equipment, were the two 'yogis' whose steps I had been dogging for the past fifty miles.

Chapter 8

A Rumour of Elephants

As I finished a short resume of my three-day trek, Dharmananda observed: "Your meeting with the king cobra is a good omen for you." He spoke with his usual good-humour. "We Hindus believe that Siva sometimes manifests Himself on Earth as king cobra. To appear before you like He did is a sign, a signal that you are on the right path. We all need these reminders from time to time and should try to recognise them when they occur."

The morning's *dharma* talk was sprinkled with several unfamiliar faces. Some had travelled many hundreds of miles across India and sought only to relax, to hang out and recover energies lost on gruelling bus and train journeys.

Dharmananda signalled his own awareness of the fact by reiterating the benefits a newcomer might expect from a few weeks' stay in his ashram, and how he saw his own role as a spiritual teacher. His eyes ranged gently about the room as he outlined his hopes:

"I am not interested in numbers; whether there are thirty here or one hundred, it does not matter. I want a small number of people who are dedicated to developing themselves." The emphasis fell heavily upon the final word. "Those who want to change the world but not themselves – I don't need them! I wish to start at the beginning, run a yoga kindergarten. But Westerners – you are all in big hurry to get enlightened. You only want *Super Yoga*, a Yoga University!" A glimmer of resignation tempered his criticism: "It is true that if I advertise 'Yoga Kindergarten' in Rishikesh no Westerner will come!"

Again the walls of the room rang with appreciation of the swami's gift for leavening serious observations with a touch of humour. He paused to gather his thoughts.

"What is Yoga? Yoga is the science of life. I teach the science of life, some Indian philosophy and religious ideas. But I am not trying to sell you Hinduism. People from all religions are welcome to stay here.

Through the regular practise of Hatha and Raja yogas, you will awaken possibilities which lie latent within each one of you. I will also introduce you to other forms of yoga such as Japa yoga, Bhakti yoga, Mantra yoga and Karma yoga." He chuckled softly. "This last one – 'Karma yoga'- meaning 'selfless work' is not very popular with Westerners. If I ask at the end of my talk for volunteers to spend only an hour on some cleaning task only eight or nine will stay behind, while forty more people – in a flash, they are all gone!"

He drew our attention to a woman of serene composure, the only nun, so far as I had noticed, to be living in the ashram on a day-to-day basis.

"Mataji – I am sure you have all seen her going about her daily tasks in the ashram? She is very quiet, says little, but is a woman of tremendous strength and character. She is a shining example of Bhakti yoga in action. For the past fifteen years she has lived in this ashram, a devoted disciple of our guruji. She comes from Punjab. Very early in life she decided not to get married, but to renounce all mundane pleasures to lead a simple life of renunciation and service to guru. Now she is completely free from all desires, attachments and cravings; she is at peace with herself and reflects that peace and calmness in all her activities. She is what spirituality is about – a source of great inspiration to many! Like our guruji, her own needs are fulfilled by tending to the needs of others. She is one of those rare examples of moral and spiritual strength, an embodiment of selfless activity."

An unexpected trace of emotion crept into the swami's voice as he revealed an unmonkly disclosure:

"I love this woman. Had I not thrown everything up for the life of a monk – then the woman I would want for a wife… would be Mataji."

He told a little story to illustrate her qualities:

"When I first came to this ashram ten years ago, I was not very patient man. I would easily get angry. One day Mataji brought me delicious soup of hot *dal*. I took hold of that bowl – and threw it! Mataji said nothing. She went away and returned with a second bowl. I threw that away also, I was so angry!" The swami's voice filled with remorse. "Only when this woman returned a third time with a bowl of *dal* did I realise the stupidity of my own actions. In a second I understood her great qualities: I fell at her feet."

He paused for a moment's reflection. "That was a great lesson for me. When we are angry, we should throw it out in a way which does no harm to others. Seeing Mataji's reaction was very instructive for me. I was angry because of the thoughts in my mind. It is like a computer: if we feed garbage in, only garbage will come out."

"Why do bad thoughts come?" asked a Frenchman sitting with his back to the wall.

The swami leant forward:

"They come because there is something corresponding inside us. When you go outside the ashram…walk in the bazaar, you enter the mental atmosphere of a public place; it is a tangle of thoughts and ideas, some of them not very nice. Not only your head but your mind, is in the middle of it. One must have a very steady sense of inner direction to live harmoniously in the world. A filter of consciousness operates, making you aware of some thoughts and not others. What governs this are your inner affinities, your inner habits and mental attitudes. Thus the nature of the thoughts you have will be an important indicator of the kind of character you possess. If you really want to progress spiritually, you must meet all that is within you which doesn't want to progress."

That night I lay on my bed and watched through half-closed eyes as the flickering candle cast strange shadows on the bare walls of my room. What is meditation? Is it just being here, in this moment, lying down...watching...being aware? Then our wandering minds pull us into yesterday...tomorrow. 'What will happen? Why did?' Our bodies are in one place whilst our minds are in another, chasing the grand delusion that only a few don't fall for: tomorrow will be better than today. Was I one of the few? I still felt a strong pull to go to Gangotri. I don't know...

Sometimes in India I felt I was simply moving deeper into the moment...towards the centre. Not moving from A to B to C from Monday to Saturday, but A...A...A...AUM...AUM... Occasionally, for an instant perhaps, we live from the centre of our being. We are the centre, the mandala's still point, before change happens and the eternal rhythm draws us away, back to the edge...we live faster. So the cycle goes and we have to go with it, live from where we are in our innermost selves. Isn't that living truthfully?

A stray waft of Himalayan air blew in through the window, causing the candle flame to duck and waver. I reached out towards the table and cupped my hand protectively about the light. When it settled I withdrew my hand. At the same time a clear thought arose in my mind, a decision.

Outside in the silvery-blue night, the ashram generator's steady *chug-a-chug* skipped a beat before fading into silence. Soon, I followed it into sleep. When I awoke in the morning I found a scorch mark on the table where my candle had burnt itself out.

The bazaar was its usual tangle of smells, sounds and colour. A steady stream of worshippers, some of whom carried flower garlands, flowed through a temple gate, past which I glimpsed a radiant garden planted with rustling, silvery trees. From beyond the boldly painted statues of benevolent deities, bells, accompanied by the rise and fall of devotional chanting, tinkled with jewel-like clarity.

A fruit-stand *wallah* opposite the gate was rearranging apples and oranges in neat formation on a wooden push-trolley. On the nearby wall hung a bundle of fresh sugar cane, tied loosely and waiting to be fed through an antiquated pressing machine. At the foot of the *ghat* where I had first spotted 'Greystoke', an old man was standing knee-deep in the Ganga, oblivious to all else, save the patient, devout splashing of the holy water over his exposed upper body and head.

It was a fine almost cloudless morning, but there had been heavy rain during the night and the air still felt pleasantly cool on my bare arms. Squatting beside his steaming kettle a white-clad café owner gave me an acknowledging nod. A small boy with wide, almond-shaped eyes crouched on a step rinsing *chai* glasses in a pail of water, his face attentive to his task. It is the same all over India. Again and again, I saw someone on the streets taking tremendous pains with the most mundane of occupations.

Perhaps part of the explanation lies in the Hindu belief that by working hard even the humblest street cleaner may climb the ladder of caste and one day reach *Nirvana*. In the big cities the caste system in its traditional form may be breaking down, but for many millions the structure – not so dissimilar to that of Medieval Europe, but still repugnant to Western minds – is a fact of existence. Each group serves the interest of its own kind; people feel bound together by a common purpose, even if it is only a life of beggary. If we dismiss *karma* as no more than another example of backward, superstitious nonsense, claiming that the real causes of such injustices to be, say, ineffective birth control programmes or political exploitation, then we display a great ignorance of our own past. For several hundred years after Christ, *Karma*, in fact, was part of the Christian faith: "Whatsoever ye sow, that ye shall also reap".

It should also be remembered that the achievement of the final Goal is a matter upon which Hindus differ intensely. What they <u>do</u> agree upon is that all people are bound by a succession of imprisonments (*Samsara*) in bodily existence, the wheel of destiny by which all human beings are chained by their desires. These desires cause attachments and suffering. Only by coming to know God are the fetters of the senses cast off.

For the Buddhist all life is suffering or *dukkha*, unsatisfactoriness, to use a less emotive term. God does not enter into the equation. The cause of suffering is seen as *avidya* (ignorance) and liberation is achieved by following the teachings of Buddha.

I quizzed Dharmananda on the subject:

"Is it not possible for, say, a bicycle repair man to save money, move to another town, and open a garage?" It was a simplistic question.

His response was emphatic:

"No, if that man did as you say and moved to another town, he would be recognised immediately. Even if he were to move many hundreds of miles, people would know him: by the way he spoke, by the way he dressed and many other details. They would know at once to which station in life that man belonged. If he tried to open a garage, he would probably be pulled into the street and beaten, in some cases even killed."

At the T-junction near *Choti-Wallah's* restaurant I swung left by a pastel-coloured temple-shrine and paused at the second of two gaudy music kiosks selling Asian tapes and Western middle-of-the-road Pop. The *hip* owner of the booth, with his Bollywood hairstyle and Zappa moustache was a model of DJ cool, whose musical interest centred exclusively on the Western pop scene. His face betrayed a touch of disdain as he admitted,

"We have traditional music – <u>if</u> you want it; but it is not modern.. Have you heard the new release by Eric Clapton?"

I had, but settled for a talk by Osho and Zakir Hussain's *Music of the Deserts*.

The bridge was already rocking when I stepped on to its tarmac surface. At the centre I stopped and peered down through the wire screen safety fence. Below, imbued with a strange intensity, the grey-blue ghostly forms hovered against the swift current.

Because of the rain, the Ganga was higher than usual, freeing sidewater jams upstream and ferrying them down in single logs or in tangles of branches, low floaters. A particularly large branch, its twigs still bearing leaves, lunged beneath the bridge. Some of the fish, disturbed by its shadow, peeled away from the main shoal, diving deep until I could no longer see them. I watched its bobbing progress downstream for a few moments before crossing to the opposite railing.

I was just in time to see another flotilla of dead branches, stumps and clumps of coarse grass surge round the foot of a high, vegetated bank; a blackened bough rose from it like a withered arm of the drowning giant which it might once have been. A few yards ahead of the main mass a solitary object floated gently on its own as though leading the way. It was obviously just another tree stem to which I initially gave no further attention. Only when it swung lazily in the current to give me a different angle of view did my gaze begin to fasten upon it. Something was different. The log, or whatever it was, bore down upon the bridge with portentous speed. I saw it was going to pass directly beneath me and held my breath as it grew swiftly larger, revealing more detail, a green too vivid to be grass.

All at once – I cannot pretend to know the exact moment when the truth struck me – a void seemed to open within me as I realised it was a body.

The woman floated face-down in the water, her limbs arranged in an odd symmetry as though she had perished while kneeling on all fours. Her village clothing of living greens and yellows was a shocking contrast to the stark fact that she was dead. While I was unable to see her face, I noted that the flesh of her upper arms was smooth and rounded, suggesting she was not of any great age. Had she slipped and fallen into the Ganga or been thrown? There was no way of knowing. But I had the feeling that the body hadn't been in the water a long time. And a chill ran through me as I saw for the second time in a few days, a grim black sentry, a crow holding watch from its perch between the woman's glistening shoulders.

I made to return to my original vantage-point, but the space was now occupied by an Indian family, including two young children.

As the corpse reappeared, a few quiet words broke from the father, directed to the two children who looked on in wide-eyed silence. I couldn't catch what he said, but it sounded simple, prayer-like, respectful. His face, and those of his children and wife showed no signs of shock or dismay. In India people generally don't shy from the fact of physical death as we do in the West. Consequently, when it confronts them in the street they are more willing to accept it as part of life's totality. Furthermore, their tradition of cremating the dead, implies a more profound view that death is not the end, but is only a step on the way: as the outer form is reduced to ashes, the higher part continues towards the state to which his or her actions during worldly life have prepared him.

The family began to move off as a gust of wind squattered across the river's surface, hiding the fish from view. The corpse, bobbing with an eerie nonchalance, dipped two or three times as it slipped with increasing speed towards the turbulent bend in the river, the crow now a ragged stick-creature, surfing bolt-upright through the tremulous waves, as if towed by some tiny invisible boat. It was time for me to move on as well.

The English couple sitting in the café a hundred yards further on looked around thirty, though the man's air of crusty impatience made him seem older. His voice, as I dropped into the tiny restaurant's only other table was curt, every word seemed to carry a repugnant whiff of dismissive finality. He made me think of school prefects, Sandhurst training and power-dressed marketing chaps. It's stereotyping, I know, but I immediately felt a warning ripple of unease in my stomach region. In such cramped circumstances conversation was inevitable sooner rather than later.

The man especially, made me feel I was drinking the waters of yesterday.

"We're here for a few weeks' trekking actually," he announced, when curiosity finally got the better of me. I nodded back, but for a while remained silent. My all-too recent experience on the bridge had left me in a thoughtful frame of mind in which conversational pleasantries with strangers from my own country was the last thing I needed.. No-one will know if you re-invent yourself in lands far from home. It's so easy until, that is, you come face to face with the unflattering mirror image of yourself. You realise you're all strangers in the subcontinent.

"And you," the Englishman said presently, "what are you doing – hanging around?"

A spark of irritation flashed through me at the barely-concealed irony. He had obviously decided about me. I glanced at his companion. Her eyes were closed as though in exhaustion.

"I'm staying in an ashram," I responded quietly, returning the metallic stare which gazed at me out of a sun-reddened face. He wasn't a head-tripper, I could see that.

Before I could continue, he cut in:

"'Ashram', what's that, some kind of..?" He looked at me as if I might be an escaped suspect and liable to fits.

"It's a place...a place you can stay, do yoga, meditation," I began, my voice tailing off as I sensed the other's suspicion. Something rose in my throat and I was suddenly aware of the harsh light bathing the wide open doorway.

The hiking man's eyes gleamed with sardonic humour,

"Oh, I see...Ju-ju, Indian rope-trick – one of those places!?"

His companion's eyes fluttered as if about to open. I tried to gauge whether she was fatigued; or merely bored.

Drinks were brought to their table and a lengthy silence followed. It was clear that my 'weird practices' were of only sideshow interest. I looked at the restaurant's owner who usually had a helpful word or two for one's plans.

"I leave soon, for Gangotri," I said.

His mouth opened to reply, but before he could do so, the Englishman interrupted: "Gangotri!?"

"You've been?"

"Tried to more like." The woman nodded into her glass of freshly squeezed orange juice. "We were on the bus," he continued tersely, as if they were both back on the trail with not a moment to lose. "The bloody thing broke down – typically Indian...decided to walk. Eighteen miles to Gangotri. There was an enormous avalanche...Road buried ten feet under. Too dangerous to continue...We turned back."

His spoke in a peremptory, headmasterish way.

"How long ago was this?" I replied. "The road crews are usually very quick to clear."

He was having none of it:

"I should abandon that idea, if you've any sense!" His eyes drilled into mine, like one who doesn't countenance dissent.

I had noted the khaki short-sleeved shirt with matching breast-pockets and the rugged walking boots and was tempted to a dry retort. But I restrained myself from such drollery and told him, "Many thanks, but I think I'll head up there anyway, to see for myself."

176

The man was an obvious case of an all-conquering ego and I felt sorry for his girlfriend if she was as unassertive as she seemed.

By mid-October I knew the so-called trekking season was fast drawing to a close and before long the higher mountain roads and passes would be blocked till the following spring. Clearly, I should not leave my departure longer than another week, otherwise there was a very good chance of God's gift to the army being right.

That I had a natural genius for collecting disquieting experiences that morning seemed to be confirmed on my way back when, re-crossing the bridge, I all-but tripped over a gloomy-looking yogi. More properly, a *fakir* or magician, he was squatting by a narrow bed of four-inch nails smoking a *beedi*. I was mildly surprised since by reputation, those in possession of amazing physical and mental powers, are widely considered to have no need for the pleasures of the common herd.

A shrewd gleam lit the man's face at the approach of another unsuspecting foreign meal ticket. But I was not in the mood to linger as a one-man audience for a few tricks and have to fork out for the experience. The genuine yogis, such as the silent cave-dweller, radiate a palpable energy which all who come near can feel but I felt nothing except a kind of greyness from this man and turned away. When I glanced back a dullness had settled once more over his malleable features.

The yogis of old understood man's natural curiosity for the unusual and miraculous, but warned that occult powers were only sidetracks, distractions from the True Path. One who set his mind on developing the *siddhis* as the powers are called, would remain stuck there, unable to rise higher.

One of Ramakrishna's stories illustrates the dangers of psychic powers misused:

Once a great Siddha (a spiritual man with psychic powers) was sitting on the sea shore when there rose a great storm. The Siddha, being greatly distressed by it, exclaimed, "Let the storm cease!" and his words were fulfilled. Just then a ship was going by at a distance with all her sails set; as the wind suddenly died away, the ship capsized, drowning all who were on board. Now the sin of causing the death of so many persons accrued to the Siddha, and for that reason he lost all his occult powers and had to suffer in purgatory.

Later that day I wrote in my journal:

"When I am too long in the company of Westerners, I begin to lose sight of India.

Monkeys. A daily nuisance. They swing unexpectedly down from the roof to snatch food from plates, even entering rooms when the opportunity arises. Lost a banana at breakfast. Indians bang a stick against a wall to scare them off.

Energy. Lying under my fan during the hottest part of the afternoon, a quiet lassitude takes over from the vitality of the morning. Monotony. Part of me still resists having less energy here. Go with it. Until sundown it needs an effort of willpower to do anything.

Mountains. Drawn to them because something inside me resonates with these high empty places. Normal limitations of perception transcended; the mind free to expand beyond everyday boundaries. At sacred places, a subtle spiritual vibration. The visitor feels his own spirit interfused with something greater. Absence of other human minds a help in this. Alone but not lonely."

One morning I set out with the Canadian, Marshall, to look for elephants. I was inspired to do so by an Australian woman's dawn encounter less than an hour's walk down river. A young helper at the ashram had also crossed paths with a large bull elephant whilst cycling along the dirt road above Ved Nikatan at night. The chances seemed good.

We walked in silence, heading left from the ashram entrance down a sandy lane dotted with protruding rocks. No-one was about and we had the way to ourselves. A faint wind blew off the Ganga, cold against my exposed hands and face, and carrying with it a jangling of temple bells from the opposite bank. Above, a silky mist still hovered, gossamer-like in the tree canopy. As we passed a lone water tower, I scanned the way ahead hoping for some sign or sound that large creatures were in the vicinity. Nothing stirred. Some of the taller trees rose like ghosts of themselves into the pearly whiteness. I spotted a walled barren space on our left which looked as if it had once been a garden.

Presently the path dwindled, giving way to isolated boulders sprouting from a raised shingle beach. At the top of the short slope we came upon a small boy and an older man warming themselves under a ragged awning where they had spent the night. A twist of smoke rose feebly from their meagre fire.

"Namaste." I forced a smile, aware that we must have taken the pair by surprise.

The man raised a broken stick in short acknowledgement of our existence, but said nothing. The boy looked chilled and undernourished.

The forest, cold and gloomy-looking, was penetrated at this point by a dry streambed which seemed to offer access to the more tangled terrain through which I had found my way to Neelkanth temple. I drew Marshall's attention to the possibility of using it.

"Yep," the Canadian agreed, after a short pause to study the way ahead, "might as well."

The skin was stretched taut across his cheekbones, but there were half-moon shadows beneath his eyes which hadn't been there before. I thought he looked hung-over, unwell, but didn't broach the subject.

The streambed walls were draped most of the way with dank foliage, but the gradient was slight and we ascended the bone-dry channel easily enough, with only an occasional halt to determine the way. An invisible bird or two screeched from the nearby trees, otherwise the tall forest was filled with a sombre, introspective silence which caused us to talk in hushed tones, like felons in some stage melodrama. Unusually for me I felt short of energy and wondered what might by the cause. However, it didn't feel fatal and I kept on as if nothing was amiss. Nevertheless, when our sortie brought us to a tell-tale pile of dried dung, I was glad for a short halt. I poked it with my toe.

"Yesterday's," Marshall grunted, as if talking was a big effort.

We moved on, sometimes together, occasionally in single file as the occasion demanded. For some reason I felt remote, detached from what we were doing, we two gaunt souls in search of a mirage, a rumour of elephants. I followed slowly behind, wrapped in my own desolate thoughts, hearing only the occasional drip-drip of water from the leaves, to break the tomb-like silence which pressed in upon us from all sides.

Presently, Marshall halted, cocking his chin towards a bush twenty yards ahead, where a narrow path made a crease in the streambed. "Hear that?"

I stopped short, conscious only of my own breathing. And then I heard it. Slow measured footfalls approached accompanied by branches being pushed aside. I held my breath as the nearest bushes trembled. Next moment the foliage parted and –

"Shit!" Marshall exhaled in disgust as a mule bearing two pannier sacks of river gravel plodded into view. His mouth curled with sardonic amusement as the creature padded across the streambed with nonchalant indifference to our rooted presence. Close behind, the mule's owner advanced into our line of vision, a grizzle-cheeked man clad in a tattered shirt, or *kirta*, and *langoti*, an abbreviated version of the *dhoti*, a garment often worn by labourers.

Despite the slower working of my brain at that hour, the fleeting encounter touched something within me, no less and no more than might have an ascetic, seated by the Ganga's side. The man nodded, but did not break stride or speak, only tweaking the beast's rump with his stick as both of them vanished from sight.

I exchanged looks with my companion:

"I think we're too late for elephants," I suggested. "We should have been around *at* dawn, rather than an hour after it."

Marshall raised his gaze to the mist-shrouded heights, "Yeah, mebbe we should've - " His expression was deadpan. The morning was still too early for either of us to show much emotion.

We had only been on the lookout for half an hour, but my instinct told me we would be wasting our time to search further. Elephants are normally wary of humans, and to have stood a chance of spotting one required being up and around before anyone else was. There was no point in our continuing. But I hesitated to be the first to suggest we return to the ashram, so I mentioned another trail I'd noticed near the old man's camp which might be worth investigating. Marshall said he'd seen it too, so we turned about and set off back the way we had come.

After a few minutes, we clambered up the side of the stream bank and angled left through the trees until we could see the river shining directly ahead.

I felt even colder close to the bank and wished I'd worn an anorak over my pullover. The footpath rose sharply up a short, vegetated wall before bringing us to a sudden viewpoint overlooking a flat-roofed stone dwelling with the Ganga sparkling directly below. With his bare back towards us a tall, *lungi*-clad sadhu stood gazing silently over the timeless scene. A line of orange washing suspended across the verandah, suggested the building was home to several holy men, though of what kind I could not be sure. Feeling like intruders we moved on without talking, lest we disturb the man's solitary observance.

Where the path might lead us I had no idea, though my hope was that it might take us to a band of precipitous ochre-tinted cliffs which lined the sweeping bend some two hundred yards down river. I was therefore not expecting it to finish so abruptly only fifty yards later, at the bottom of a flight of some fifty stone steps. A hefty-looking door set in an old brick wall at the top aroused my curiosity and I started up to investigate.

The steps, themselves encroached upon by creeping plant growth, were the only access to the foot of the wall, the rest of the slope being a tangle of bushes and thorns. A line of rusty wire protected the top of the wall, which was about eight feet in height. I aired the view that it might be a military compound of some kind.

Marshall shook his head, "It's not that," giving me the impression that it wasn't his first visit.

As we neared the door I began to feel an overwhelming sense of loneliness emanating from the decaying, worn-out walls. From within arose a mournful silence. I tried the door with my shoulder but to no effect; whoever had locked it had intended to keep intruders out.

We stood for a few moments before the barrier which, for all we knew, might be guarding a temple or some ruins of archaeological interest. I was still keen to investigate further, and by standing back from the entrance and drawing myself up to my full height, I was able to see the tip of a small building – a pale orange dome, peeping above the top of the wall. I had the feeling that this structure was of more recent date than the time-eroded wall which held us from it.

I drew the Canadian's attention to the possibility of crossing the wall by climbing a convenient tree.

He dismissed the idea with a wave of his hand: "If we go back an' round right – there's a gate, I remember."

"You've been here before?" I said, as we abandoned our position and began to descend.

Marshall was already a few yards ahead and appeared not to hear me.

However, he was as good as his word and shortly we came up to a gatehouse manned, so to speak, by a gently-snoring guard, lying full-length under a blanket. He did not stir from the earth floor as we trod silently past him to enter the enclosure beyond. I was filled with great expectations, yet at the same time aware that we ran the risk of being challenged at any moment.

At this point Marshall dropped his curtain of early morning rectitude and became comparatively garrulous. "Know whose place this used to be?" he quipped from the corner of his mouth, as though we were in a detective movie.

A flash of thought transference seemed to dart between us.

"Maharishi...TM," I shot back.

"This was the place all right...The Beatles were here in the 60s." It was hard to believe.

I had had first-hand experience of the popular yogi's meditation methods when I lived in Vancouver. At the time, he was an engaging cult figure from the East whose beatific expression, robes and beads, not to mention the chance to develop the *siddhis* or 'higher powers', attracted many followers from the psychedelic generation in search of ecstatic experiences. Along with others, I received my so-called personal sound or mantra, the constant chanting of which would , we were told, bring us finally to a state of higher consciousness, pure Bliss. Little if anything was mentioned of the method's roots in ancient spiritual traditions. It was a bespoke technique stripped of most of its Oriental mystique and neatly packaged for Western consumption. Even the army and business management courses were offering it, as a form of stress control.

For a while my head was like a thousand-watt electric light bulb, an incandescent, all-knowing globe of energy. I felt capable of great things. If I really wanted to I could even fly! IF. In a month or two the highs became fewer. I began to have doubts. Others were using my mantra I discovered, and I felt uncomfortable around the clean-cut, conservative-suited facilitators with their ready-made, pre-packaged moralistic answers to controversial questions. I moved on. But the seed had been sown and now – years later, its life-force had brought me to India. A Western education was not the ideal preparation since, to give but one example, the intellect cannot cope with paradox.

When questioned about his wealth, the Maharishi replied, "I like money – there is nothing wrong with it. It does not interfere with meditation".

181

I could see he had lavished some of it in creating his garden-ashram in its peaceful location by the Ganga.

Before us, the ground rose in a series of swellings or Tolkein-like mounds. Trees carefully located to please the eye over thirty years before, now stood in wan neglect on the rounded slopes. A narrow footpath wound a circuitous route amongst these miniature hills, disappearing behind one, only to re-emerge further on, carrying my interest with it. The most eye-catching feature however, were not the hillocks – but the elfin buildings whose designs seemed to owe much to hallucinatory experience.

Ever-alert to the distinct possibility that we should not be there, we proceeded cautiously up the circling footpath. What the Maharishi seemed to have had in mind was to create a kind of garden-village for himself and his followers, an oasis of everlasting summer, a holiday camp of peace and love away from the hustle of the outside world. Had the sun broken through the shroud-grey mist at that moment, I might have warmed more to his fairy tale vision than I did. But the absolute silence and the absence of people was eerie, to say the least. The place struck me as being more like an abandoned film-set, waiting sadly for the stars to return and the cameras to roll. An intriguing Never-never Land.

Marshall kept his thoughts mainly to himself, but I'm sure he had impressions similar to my own.

There were more of these unusually striking creations than the view from below indicated; more than I could be bothered to count. Eight feet or so in diameter, they were sprinkled liberally along the margins of the footpath as though they had sprung up overnight like a lane of giant magic mushrooms. Each was set in its own miniature plot and of perfect onion shape about the height of an average person. Skilful brickwork might have been used or they could have been cast in cement – it was impossible to say. Tiny windows, deeply set in the thick, pastel-tinted walls and the dolls-house doors were expertly crafted from good-quality timber still in sound condition. The doorways were so low that only a child or dwarf could pass under the lintel without stooping. Were they dwellings, meditation 'pods', or what? With a minimum of worldly goods they might have accommodated two at a pinch. Or perhaps the old wizard had a surprise up his sleeve and the mushroom-onions really communicated with a carved-out subterranean magic world of mediation and Pop-marriage of East and West. The doors were all locked as if at a spiritual Butlins during the close season, while the windows admitted insufficient light for me to make more than a shadowy guess as to what lay within. And yet the place was so obviously being maintained despite its gloomy atmosphere.

Thirty or more years on it was hard to imagine that the Beatles, 'the world's biggest band' had once sat at the feet of the benign yogi in this very enclosure. But if, according to Dharmananda, the Maharishi had not been near the place in ten years, why was it being kept in such a good state of repair? Perhaps he hoped to return one day with his devotees when TM once more became fashionable? His beard was turning white in the 60s so he's getting on.

Meanwhile, Marshall, who liked to keep to his own pace had forged on ahead, slow for him, but still managing to convey the impression that a free meal awaited whoever was first to complete ten laps of the ashram-garden. I used to be that way too, always heading somewhere fast.

Finding myself within the yogi's abandoned ashram, reminded me of an occasion in the mountains of north-west Scotland one summer. There was no train or bus, so I was forced to hitch-hike. After two hours standing on a lonely road, I was still without a lift. It began to rain. I started to unfasten the flap of my rucksack for my waterproofs. A strange thing then happened which I have never forgotten to this day. Although it was some years since I had lost interest in TM, I still found some profit in reading his occasional book. One of these, Seven States of Consciousness, was lying face uppermost on top of my belongings.

As I raised the flap the Maharishi's face, which was looking up at me from the front cover, gave me a broad wink as though to say, "All's well". Astonished, I repeated the action, raising the flap several times. But it never happened again. I had drunk nothing except water all day and was at a loss to explain how it happened. I can think of several rational explanations, but I am sure the doubting reader will already have thought of them for himself.

A low whistle jolted me back into the 90s world of sober reality. The Canadian was standing by a fence fifty yards higher and had obviously seen something. He signalled for me to get a move on. Following the direction of his gaze I saw that an obstacle now barred our intentions of further illicit wanderings – another guard-hut, with a motionless figure crouching inside. Hearing our approach, the guard, a moustached soldier muffled in greatcoat and scarf, straightened up from the embers over which he had been warming himself. He greeted us quizzically as well he might, his glance falling to the lower gate, where he would reasonably have expected us to be turned away. But if he was put-out by our sudden appearance, he didn't show it.

"*Namaste*," I greeted him, confident that a bold approach would see us through.

Encouraged by his response, I swept the upper enclosure with my arm: "Maharishi?"

The guard nodded.

"Possible we take walk round…ten minutes?"

The moustache twitched: "No sir, not possible."

I considered the wisdom of offering the man a few *backsheesh*; he looked perished, as though he had been on duty the whole night. I exchanged looks with my companion. The guard's face was not unsympathetic but I didn't wish to push our luck. All of a sudden I felt weary to the hilt of the Maharishi, the Beatles and the whole nostalgia thing of the stoned 60s. I was tired of elephant hunting too. I hadn't eaten or drunk anything since rising and was beginning to feel drained and hungry.

Marshall, who had joined a Hindi language class in the bazaar, said, "*Dhanyawad*," as we turned away.

At the outer gate, the other guard was just stirring from his blanket as we passed him for the second time. His eyes interrogated us with mild surprise, but he said nothing, allowing us to proceed unchallenged.

Beneath the trees Marshall's face looked pale and drawn. "I think I'll head back to my room for an extra hour," he drawled softly, staring ahead.

As we neared the ashram, however, his voice became stronger. Turning towards me he commented, "Y'know, you've got a few years on me, but aerobically, I'd say you're fitter than I am."

"Maybe," I said, "the *asanas* help a lot in that respect."

Remembering his remarks about the ashram's supposed lack of activity, I directed our parting exchange upon that:

"Perhaps you should get your mail re-directed up to Neelkanth temple. You could collect it each morning – three hours up, two down: you said you needed more exercise!"

He laughed easily as we went our separate ways.

I made one more lone day-trek before leaving Ved Nikatan. Kanjapuri, another hilltop temple, which I had spotted from the back road to Laxman Jhula, was the highest point on a six thousand-foot forested ridge filling the horizon above the far bank of the Ganga. Its shining dome was like a pearl rising from a sea of green.

I have long-since abandoned the hackneyed myth that the best viewpoint on a mountain is from its summit, but the temple afforded a peerless, birds-eye prospect when I reached its gleaming walls after a walk of some five hours.

The *stupa*-like edifice, encircled by smooth terraced walls, commanded the highest lookout point for miles around. Forested ridges, here and there broken by high rock walls, deep, sea-green valleys – wherever I glanced was another spectacular vista of the Garhwal Himalayas. And crowning the very furthest, film-like ridge, some two days' bus ride away, the world's highest mountain range reared up in a culmination of sharp-tipped snow peaks and dizzy arêtes. Gangotri – the end of the road – I felt a longing to be there.

Two days later I set out.

Marshall was standing on the ashram steps staring with rapt attention across the river. His profile, picked out against the sun, seemed to be looking into the void of an important self-discovery, as if his being was beginning to move to a different drumbeat, only he hadn't quite caught the rhythm. Not yet.

I said a few mundane words: "That Argentinian friend of yours – she seemed upset when I said goodbye just now…Strange, I only spoke to her a couple of times."

"Yeah?" After an interminable pause the Canadian reeled in his vision, while struggling to focus on what I was saying. A reflective gleam enlivened his eyes: "Yeah, we were talking…She said you were a peaceful guy."

I followed his gaze out over the sheen of water. On the opposite bank, timeless anonymous figures dipped and rose in fluid motion by the water's edge. Was I doing the right thing in leaving? Tao Te Ching wrote: "Without going outside you may know the whole world…The further you go the less you know".

'Not clinging', I let the idea drop, at which point my own phantom returned, frayed at the edges but intact.

Marshall took my address. He was heading for Australia to visit his brother, he said, adding: "I wouldn't mind doing some climbs in England on my way back to Canada." He gave a short laugh. "Next time we meet, we'll both be talkin' *Hindi*!"

I never heard; and I often wondered why. Perhaps I should have checked <u>his</u> room?

On the trudge over the bridge to the bus station I picked up a scrap of butterfly wing from the dust. I still have it.

Chapter 9

Caves and Snow

Rigid itineraries and over-planning have never had much appeal to me while travelling: I tend to follow the urge to set out and let the next hotel take care of itself. I am in favour of early starts, on the other hand.

With the driver's hand hard down upon the horn, we ploughed a non-stop furrow through the blue haze of Rishikesh's main thoroughfare, and were soon taking the sharp left-hander by Sivananders's ashram. Ahead, rose the tortuous climb up the range of hills capped by Kunjapuri temple which I had visited a few days earlier. Following some anatomical jarring on other buses, I had taken the trouble to secure a seat about halfway between the two axles, so as to minimise the jarring from the many pot-holes we would surely encounter before reaching Uttarkashi at the end of the day. Savouring the prospect of a change of surroundings, I settled back in my seat to enjoy the ride.

At first the views were restricted by the screen of roadside forest and the endless corkscrewing bends; but little by little the drops below my window grew steadily more perpendicular and thought-provoking. The terrain began to open up like a relief map without boundaries.

Roadside wrecks are a common sight in India, a land in which driving is only theoretically on the left, but when the back of another truck was spotted in a tree fifty feet below the road, I craned my neck to see, like everyone else. Such reminders of the briefness of life are a recurring feature of Indian travel. Some 50,000 people are killed annually on her roads, more than in the United States and with only a fraction of the number of vehicles. Ten miles farther on, near the colourful village of Narenda Nagar, we passed another road victim, a dead buffalo. Lying upside-down at the road's edge with stiffly-splayed legs and ballooning body, it looked as if at any moment it would take to the air or explode. Indian people are used to such spectacles and the only one to take much notice, was myself.

Three hours north of Rishikesh we began to descend towards a flat valley floor and the township to which I referred earlier, namely Tehri. The final mile of the approach found us lumbering across a large-scale civil engineering 'development'. It was an ashen landscape of chaotic gradients up which enormous dumper trucks growled like crawling lunar vehicles; trees uprooted and crushed in mangled disarray lay tossed aside on piles of quarried boulders. Dust hung thick and choking in the air, an orange-grey miasma which obscured the hills from view and coated the half-naked bodies of the sweating labourers who toiled within it. A guide book dismissed Tehri as 'grubby and insanitary', but following so closely after such a disheartening spectacle, I found it more like a holiday town in comparison.

Many ancient tribes such as the Bhutias, Doms, the Jads and the Ban Rajis live in the Garhwal region. Most have gatherings, *rangbangs*, where the younger people meet in order to find a marriage partner. Another common feature, polyandry, is still acceptable practice amongst these mountain people. The township has a colourful history as an ancient capital and boasts numerous temples and palaces built by the rulers of long ago. Four Maharajas have ruled from Tehri; the icecream-coloured palace of a more recent one, the former Maharaja of Tehri-Garhwal overlooks the town which still retains vestiges of its romantic past. On a hill above the square stands a Gothic clock tower built by the last ruler, Kirti Sah, to commemorate Queen Victoria's Diamond Jubilee in 1897. Every region had its army and the district is home to the famous Garhwal Rifles who fought in the Nepalese war of 1814-15. After the Treaty of Sagauli in 1816 the Nepalese left and the British moved in.

Standing at an elevation of 2326 feet, Tehri is surrounded by well cultivated terraces. It is also at the junction of four important roads which turn off to Devaprayag, Uttarkashi, Srinagar and Tilwara. Stone sculpture used to be widely practised, but today only wood carving continues on any scale; many examples are to be seen in temples across Garhwal as well as in the shops in the popular centres of tourism and pilgrimage. The populace know many folk songs composed to celebrate special occasions. *Chhura* is one which is sung amongst shepherds and takes the form of advice given by elders to the young, they themselves having learnt it whilst tending their herds of sheep and goats amongst the bare mountain pastures. It is a land of popular myths, many of them connected with the caves in which *rishis* dwelt centuries ago. Not far from Badrinath is Vyas Gupha, the cave in which *Rishi* Ved Vyas is said to have composed the story of the *Mahabharata* containing the immortal words of Krishna and the struggles of Arjuna.

Beyond Tehri, the Bhagirathi valley soon narrowed, the adjacent mountainsides pressed closer on either hand, and soon we were once more amongst tortuous hairpin bends and the harsh grinding of low gears. Steep narrow valleys stretched away from the road on both sides; on their slatey ridges stood solitary pine trees, along with another species with unusual lollypop-like tops. "Mountains of shale held together by soil", was how Marshall, my elephant-hunting companion described these hills – and he wasn't far wrong. On several occasions we passed stark evidence of rock avalanches and landslides; fortunately, the most recent had been some days before and we proceeded without hindrance. By now the road, river and gloomy gorge were close companions; streams plunging down steep gullies of black, shiny rock added to the drama of the scene. For a while, I followed the line of a threadlike track which clung to the opposite side of the ravine. Sometimes it passed a walled cave; once or twice it reached a blank expanse of vertical cliff-face – and disappeared altogether, only to stutter back to life thirty yards further on. It looked like an ancient *Yatra* route, but I saw not a soul on it and I had the feeling that few travellers used it in present times.

We reached Uttarkashi, 'the temple town', about 6 p.m. just as darkness was falling. Lights were going on in the streets and twinkling on the surrounding hillsides as I made my way through the softly-lit bazaar to *Hotel Meghdoot* which I had stayed at in 1991. The price of a single room had scarcely risen at all in five years: for Rs50 (about one pound) I obtained a narrow room with a single bed, a tiled bathroom and a fan.

My first surprise of the evening occurred just as I walked into the room. There was a *blip* from the hotel lights and I suddenly found myself in total darkness. In fact the entire town had been shut off by a power failure. Fortunately, I had packed a candle in my rucksack, which soon re-established a semblance of life in my drab little room. Bus-travel weary, I flopped onto the bed for a thirty minute blackout of my own. One advantage of the power failure, at least as far as I was concerned, was that it pulled the plug on a mega-volume TV set blaring from an open window across the back lane. When I eventually stumbled into the street to look for a restaurant, the shops and stalls were aglow with a colourful assortment of lanterns and candles – clearly, someone was making a comfortable profit.

Stretching along both banks of the Bhagirathi, this busy town of 12,000 inhabitants is the capital of the Westernmost district of Uttarakhand which fringes the Indian border with Tibet. Siva is the presiding deity and is worshipped round the clock in the ancient temple of Lord Vishwanath. At daybreak and dusk, the temples and shrines are alive with the mingled cadences of bells and the chanting *pandits*. One of the most important festivals is *Makar Sankranti*, on which day the streets throng with gaily-dressed men, women and children, while vividly-painted images of the various gods and goddesses are paraded through the town. Music charges the air and the day culminates in much song and dance.

After the ascent of Everest in 1953, the then Prime Minister Nehru instructed that a mountaineering training institute should be located in Uttarkashi, a town through which many pilgrims pass en route to higher, more distant holy places.

A seven-hour bus journey hadn't left me with much of an appetite but after a short sightseeing stroll, I dropped into the welcome warmth of a small restaurant

188

on the main street. I ate a snack-sized meal of potatoes and spinach, rice and butter *nan*. From a nearby stall I bought some bananas and oranges for the morning, after which I wandered back to my room and was in bed shortly after 8.30 p.m. I rose at daybreak feeling sluggish in mind and limb; a mild visitation of diarrhea and a late resuscitation of the shrill-volumed TV had left me with a disturbed night's sleep. In trying to leave early to catch the 7 a.m. Gangotri bus, I found the hotel's front door securely locked and had to rouse the reception clerk, a young man sleeping stretched-out behind the counter.

A flexible approach to timetables was definitely the order of the morning, and I decided to forego the first bus and seek a leisurely breakfast instead. Another bus was leaving within the hour, so there was no great urgency to hasten from the town. On the other hand, although I knew that Uttarkashi boasted some interesting ashrams and was the centre for a variety of trekking excursions, there was no telling how long I would be lucky with the weather. Accordingly, I dived into the first restaurant, which appeared to offer an appetising breakfast menu.

Fifteen minutes early though I was at the bus stand, I still wasn't early enough: every seat on the bus was already taken, with an arm, a shoulder or a dark face pressed to each window as if to emphasise the fact. There was only one thing for it: I took to the roof-rack amongst the bedrolls and tin trunks.

It was a fine, airy perch for a morning start and I was soon joined by four young soldiers. We left promptly, itself a surprise.

The way at first passed through a varied landscape of wooded hills getting higher, sparkling water and clusters of dwellings set amongst fields of ripening crops. We climbed past the turning to a famous beauty spot, Dodital Lake which stands amidst a thick forest of oak, pine, deodar and rhododendron at an altitude of nearly 10,000 feet.

Roof-riding required constant awareness of approaching hazards such as telegraph wires and trailing branches. On one occasion we flattened ourselves when the bus passed under a rock canopy blasted from the cliff face. After an hour of such progress the bus emptied sufficiently for me to claim a seat inside. It was a very different experience since, apart from the proximity of human bodies, the passing scene was invisible unless I forced open the smeary window. Moreover, at every lurch and pothole, the interior panelling shook and jolted as though held together by elastic. If it rained I imagined that waterproofs would be essential.

In front of me sat an alert-looking Indian with a short bushy beard and an army-green forage cap perched nonchalantly on a thatch of black hair. He drew upon an occasional pungent roll-up, the smoke from which encircled his head before trailing with increasing speed out of the window. In the adjacent seat across the aisle sat a Frenchman, who later told me he was living in America and had decided to celebrate his 50ᵗʰ birthday by visiting the Himalayas for the first time.

After a mile or two the Indian turned and introduced himself. His beard made him look much older that his years, for I was surprised to learn that he was only 24. He had, he claimed, been a member of the Bengali Mount Everest expedition in 1991. Now, having just completed another expedition, he was having a few days "trekking around" on the glaciers above Gangotri. It seemed that we were heading in the same direction.

"I want to climb three mountains outside India," he continued, "Mount McKinley in Alaska, Cerro Aconcagua in the Andes and the Matterhorn in Europe. All wonderful mountains!"

When I told him I had climbed the last-mentioned peak, he regarded me with increased interest:

"We can walk together up Gangotri Glacier…very good ice climbing!"

"I'll see," I responded noncommittally, not wishing to jump into any instant partnerships I might regret later. Changing the subject, I asked if there were any British writers of mountain literature whom he particularly liked.

"Peter Boardman," he answered without hesitation. "His writing is very poetic. Also Indian people like very much Tilman, Shipton and Smythe – you know Frank Smythe?"

"Nanda Devi…Valley of Flowers," I said.

He smiled in reflective silence and began to twirl another roll-up.

At this point the Frenchman suddenly intervened and began to complain about his smoking: "It is bad for the environment; you should give up these 'abits monsieur!" The front of the bus rang loud with his frosty recriminations.

The Indian bore the criticisms affably enough but refused to comply, pointing out quietly how rapidly the smoke vanished through the window.

According to the Frenchman, that was the whole point – its polluting effects would ultimately be no less.

The argument turned to and fro for some minutes and I began to grow weary of the carpings of a true zealot. In addition, if the Indian couldn't detect the rising tone of patronising irony – I, as another European, certainly could!

"Excuse me," I interrupted, leaning forward to make sure the Frenchman heard. "Isn't your country, France, making rather a lot of smoke in the South Pacific, these days?" I tried not to mirror his aggression.

He turned his face towards me over the top of his seat. "Oui, that is true, but in America it is not possible to smoke on the buses; it is not allowed. You are probably a smoker."

I assured him to the contrary. "When we're in other countries we may have to put up with the way things are instead of getting upset," I added, in a vain attempt to persuade him.

He eyed me sceptically before ending the exchange with a piece of irrefutable logic: "You can choose to sheet on the bus, monsieur, or you can choose not to."

Before long we began to climb even more steeply; an interminable sequence of grinding s-bends carried us to the highest point before Gangotri, at an altitude of 9000 feet. The route had risen 2000 feet in 9 miles and I was glad for a short leg-stretch at the little hamlet of Sukhi.

The views were getting more lofty with each mile and I noted an abundance of fruit trees in the vicinity. An equally spectacular descent awaited us on the other side. With extensive views of the wandering Bhagirathi below, the road ran parallel with its banks for several miles, the huge deodars slowly gave way to aromatic pines and I began to look forward to leaving the bus and walking amongst them. The opportunity to do so occurred sooner than I expected.

Scarcely had the bus began to wind tortuously downwards a short distance beyond Jhala, when we came to an abrupt halt. Ahead, the road was blocked by a landslide. There was no way for a vehicle to pass round or over it.

Along with most of the other passengers, I alighted from the bus and strolled forward a hundred yards to inspect the scene. The road was buried to a depth of three or four feet in places and one or two of the boulders were as tall as my shoulder: the immediate prospects didn't look promising.

"How long will it be?" I asked of an army officer, who seemed to be directing clearance operations.

"No problem," came the bright response, as though we were clearing gravel from a cricket pitch. "Two hours, sir!"

Just then, a voice shouted from amongst the largest rocks where, for the first time I noticed a length of trailing wire.

At this the soldier turned to the gathered assembly and spoke briskly in Hindi, waving us away as he did so. Everyone fell back under his instructions to the vicinity of the bus, myself included. I had knowledge from mining just how far rocks could be hurled by an explosion and took up a prudent position behind the back of the bus.

We didn't have long to wait.

A low *car-uumph*! And a belch of smoke started from the rubble as the charge was detonated. A split second later, an invisible projectile fizzed over the bus and whacked into the hillside behind us. No-one appeared to notice or showed the slightest concern at the near-miss.

I glanced up at the bus roof where my rucksack lay padlocked to the low railing. The surroundings were becoming more inspiring with every mile, but I had no desire to hang about for two hours more waiting for the road to be cleared. Besides, I knew that an Indian 'two hours' could well mean four.

I decided to retrieve my bag and walk on. The scenery was definitely changing, with bare summits reaching into the sky above the last straggling trees. There was a tingle of higher, as yet invisible mountains in the air; the foothills were behind us now and I felt a powerful desire to walk the last miles to Gangotri as the pilgrims of old would have done.

Rather than follow the road's lengthy windings beyond the rockfall, I took a short cut, descending steeply down the mountainside towards the river, a way which had little to recommend it, but at least I was free of the cramped confines of the bus and could smell the fragrant forest about me.

Twenty minutes later I rejoined the road in the valley bottom, paused to eat a couple of bananas, and carried on. The river at this point, wandering peacefully over gravel beds, was spanned by a sturdy box-girder bridge. On the far side I soon came up to a kilometre-stone which said:

'Gangotri 29km.'

Although it was now midday, the tree-fringed road gave regular shade, whilst the proximity of the river on my left contributed a psychological cooling effect to make the walk a pleasant one. The rockfall had temporarily halted all traffic, so I counted myself fortunate to have an empty road before me. Empty but not quite, for in a mile or so I caught up with a lone sadhu whose turban and other garments were of weather-bleached orange cotton. I thought I was travelling reasonably light, but I had to concede that his shoulder bag and metal water pot carried the candle as far as trusting to providence was concerned.

We walked side by side for a little way, but since he knew only a few words of English, I was only able to establish that he was visiting holy places in the Himalayas and had been walking for three months. When we came to the sunlit hamlet of Harsil we parted company with a wave, he walking on at an energetic pace, whilst I paused for a drink and to readjust my bootlaces.

Over a century ago a British settler, George Wilson, fell in love with this simple place, put down roots and planted a large number of apple trees. Today his orchards still testify to his lifelong affection for the locality. In addition to apples, wild apricot, walnut and chestnut trees grow in the area.

As I rested briefly in a cosy tea-stall close to the road, a tall, bearded figure came striding into view about a hundred yards away. Even at that distance I could tell by the briskness of his stride and the square-shouldered set of his upper body, that he must be a Westerner. As he drew closer I saw he was a full-bearded, weather-bronzed mountaineer carrying a three quarter-weight climbing rope of blue Perlon coiled across his shoulders. The top of his face was hidden by the brim of

Passing through Harsil

a bush hat tugged to a rakish angle; nevertheless, he bore a marked resemblance to photographs I had seen of Joseph Conrad. A pair of crampons jingled on top of his rucksack.

I thought he might give me a nod or even stop for a couple of minutes as travellers usually do when their paths cross, but his jaw was lifted, set towards some distant goal and he strode past the dingy tea hut looking neither right nor left. Seeing nothing of his immediate surroundings his gaze was locked upon the road before him as if he must get somewhere fast – wherever it was. Before an old man could hobble across the street, the crunch of his heavy, stiff-soled boots had faded and he was gone from view.

My own way led in the opposite direction, higher into the mountains. Beyond Harsil, the Bhagirathi ravine was much narrower than at the bridge, shade-giving deodars had reappeared along its course and the river itself was flowing with increasing turbulence. Ahead, the dramatic gorge cut through the granite by the Bhagirathi gave me my first glimpse of the sparkling snow-peaks around Gangotri.

The road twisted and turned as it spiralled up the mountainside. Before long I found myself looking into dizzying gullies with the river several hundred feet below, boiling between the impending walls of the narrowing, sunlit canyon. And everywhere, the ubiquitous conifers clung with silent tenacity to the most precarious of ledges – I began to sense the unique power of the Himalayas. High above the road water cascaded down dark, inaccessible ravines or drifted veil-like over huge ochre walls, the deep green of the pinewoods balanced by the brilliant sparkle of snow and ice peaks soaring into the flawless blue.

Below Bhaironghati 8 miles from Harsil, I came to the significant tributary of the Bhagirathi, the Jadh Ganga or Jahnva whose blue waters plunged through an awe-inspiring gorge.

Until 1984, access beyond this point was only by foot, the pilgrim had to descend to the Jahnva by steep rocks, cross it via a swaying rope bridge and regain the original height by an airy scramble up the other side. Now that a box-girder bridge spans the yawning gulf the road can become very busy between the months of May and early September.

Peering into the chasm, I spotted several rotting timbers protruding from clefts; they were the sole remains of the old wooden bridge which had preceded the modern one. Further down still were a couple of stick-like posts which might once have formed part of a handrail to the rope bridge. Even to someone like myself who was used to scrambling about steep rocks it looked dodgy; to visitors from the cities who had never climbed higher that the roof of a bus, it must have been a terrifying experience.

Above the next bend after the bridge I decided to take a short rest and sat down amongst the reclining granite slabs which overlooked the steep-walled gorge. The river was far below now and after eating my last banana, I couldn't resist the juvenile urge to toss a stone over the convex lip of the slab. The sport is called 'boulder trundling' in Britain and is supposed to take place during the earlier, immature stages of one's life, and preferably when nobody else is below. Such grown-up thoughts were far from my mind, however, as the stone thrummed an unstoppable arc into the empty air. My ear followed its clattering progress down the wide flaring walls. A long silence followed. I strained but could hear only a distant thundering. It <u>was</u> a long way down.

As I repacked to continue, the perfect stillness of the scene was rudely shattered by an unexpected blast from a horn. Seconds later an army jeep containing two soldiers staring straight ahead shot by. Presently a taxi, overloaded beyond belief, veered round the nearest bend, the face of its driver lit by a grin of manic intensity and the limbs of its passengers protruding from every window. I began to stagger up to the road, knowing that they would be at Gangotri two hours before I would.

Hardly had I regained it when the bus I had abandoned miles back overtook me, its huge wheels churning clouds of dust in its wake. A perspiring, bilious face was craning from a window and I thanked God it wasn't me.

194

The Ganges near Bhaironghati

An occasional noticeboard cautioning foreigners not to stray more than 100 metres from the road, warned me that I was now in a border-sensitive zone. Tibet was only an inch away on my map, but glancing around at the terrain it was difficult to imagine how such a directive could be enforced. Shortly after reading one of these signs I came to an army post comprising several tin-roofed huts amongst the trees and a stone compound surrounded by wire fencing. The only sign of activity was four or five soldiers playing a languid game of volleyball on a makeshift court. One of them spotted me and waved, shouting something I didn't catch. I waved back, pointed my forefinger up the road and kept going.

I was now at an altitude of nearly 9000 feet; the hairpin bends cut into the mountainside were beginning to unravel themselves as the angle of the road levelled off. The Bhagirathi ravine opened out into a wide valley flanked by knife-edged ridges and bristling snow-peaks. During the season the road is maintained in a reasonable state of repair, but as the night-frosts of October become more numerous, so the volume of visitor-bearing traffic dwindles. Fallen stones, unless they cause a serious obstruction, are often left where they land and vehicles have to steer round or over them.

A number of shattered trees both above and below the road testified to the destructive force of Nature. At one place on a bend, stood a huge boulder, an angular block as high as a living room ceiling and weighing many tons, its foot buried in the road's surface. On the steep slope above lay a conifer, its once-proud trunk nearly three feet across smashed in two by the rock as it careered down from the frosty heights. I was surprised at the number of stones littering the way and kept a wary ear open for any warning sounds from above.

The sun was preparing to sink below the nearest snow-capped ridges when I at last reached Gangotri.

I had heard how crowded such pilgrim centres could be and for the final mile or two had mentally prepared myself for a village choked with seasonal visitors paying very unspiritual prices. But I need not have concerned myself.

Although the road narrowed to a bumpy lane, lined with a motley assortment of vehicles (including my bus), and a row of foodstalls, makeshift restaurants and trinket shops, of the crowds I had half-feared, there was no sign. The place had an end of season tranquillity which appealed to me. Stall holders sat in their stalls whilst young boys wiped dust from already clean tables; some shutters were already up, their owners gone to Uttarkashi and beyond until the following May, when the temple reopened. Several large crows wafted speculatively over the scene. The only sound was from the Ganga, rushing through a narrow channel 100 feet below the bazaar.

Gangotri stands at an altitude of 10,000 feet in a V-shaped valley filled with trees, its name meaning 'Ganga turned north'. Tiny ashrams, tin-roofed huts and shacks lined both banks of the river with here and there an orange flag marking the position of some hermit's dwelling – a cave as often as not, cosified by the addition of neat wall-building.

The famous temple, a square, multi-domed building is situated on the left bank of the river and commands a majestic view of the upstream gorge and some of the peaks beyond, which overlook Gaumukh, the true source of the Ganga. Each year, like its sister shrine in Yamunotri, Gangotri opens on the day of *Akshaya-Tritya*, during the final week of April or the first week of May, when a *puja* is offered to Gangaji, both inside the temple and on the riverbanks. The *pujaris* are Brahmins who are chosen each year from the village of Mukhwa to deal with all activities involving the temple. In early November, the *pujaris* perform a closing ritual on the day of Diwali, or 'Festival of Lights', when its silver doors are secured for the winter and, except for a few dedicated yogis, the village stands deserted and snowbound for nearly six months.

I passed a line of low, smoke-smeared canvas shelters as I made my way slowly down the inclined lane towards the river, the ragged constructions being homes to an assortment of holy men. Where the lane divided was an open-fronted restaurant with the Western-dressed owner staring impassively across its empty tables. Standing nearby with a rucksack on his back and engaged in animated conversation with a man wearing jodhpur-like trousers, was the Bengali mountaineer.

Another heavy rucksack stood on the ground between them, and as I approached, the second man swung it on to his shoulders and made as if to be away. But I sensed that something was amiss. Seeing me, the Bengali's frown disappeared. He smiled through a robust pair of stylish sunglasses:

"I am trying to hire porter to carry my bag to Gangotri Glacier," he explained.

I thought the would-be load carrier <u>looked</u> strong enough.

"Why, is there something wrong?"

The Bengali lowered his voice, "This man no good – unreliable. I get someone else."

Another, an older man who had been following the proceedings with keen interest presented himself. The dismissed porter took off the rucksack with an expression of weary resignation and walked away. This time all was well.

The Bengali turned to me:

"It is fixed. We can eat something here – and then we can go!"

I thought the emphasis on the collective 'we' sounded ominous, so as the three of us entered the gloomy interior of the restaurant I thought I had better make my own position clear. Maybe I was getting soft, but I had just walked twenty miles uphill in an afternoon and wasn't ready to face another twelve., probably half of it in darkness across glacial moraine. Then there was the altitude to consider –

"I stay one night Gangotri – Gaumukh tomorrow," I said.

The Bengali appeared not to hear. "Shivling peak very beautiful, very difficult mountain."

"Yeah, Shivling!" I couldn't help but warm to the man's dedication.

"First climbed in 1974 by Indian border police team," the Bengali enthused. "Japanese, also British – Doug Scott and two Australians, they also reach summit by very difficult routes."

Any mountain over 21,000 feet would be no picnic to climb, but I knew from what I had heard and read, that this particular peak rated very highly amongst mountaineers the world over. Some two hundred miles north of Delhi, this granite 'Matterhorn' stands above a traditional trading route to Tibet and has attracted repeated interest from the very earliest days of Himalayan exploration.

"A great challenge," I agreed, "a lot of hard rock climbing."

I had not climbed it myself and had no intention of pretending to have done so, but I was totally surprised by the way my observation was misconstrued.

The Bengali pulled on his cigarette, his eyes narrowing behind his sunglasses. "You have climbed Shivling north face - " the words were a statement not the question of a doubter – "then I bow to you."

His dignified accolade found embarrassed ears. "No, no!" I began to protest, but I could tell his mind was made up.

"What is your name?" he persisted, obviously convinced that I was a climbing mega-star come to zap a few unclimbed peaks on a can of beans and no ice-axe. His chin lifted thoughtfully as I disclosed my identity.

"'Clark', I think I have heard," he opined knowingly, staring into the middle distance as if a Himalayan rogues gallery floated before his bearded visage.

We ate the rest of our meal in silence; the other man was obviously in a hurry to continue. I knew he still wanted me to head up the glacier with him and wasn't surprised when he raised the subject for the second time:

"Gaumukh?" he glanced at me persuasively, pushing the remains of his food to one side.

"I'll see you up there, maybe," I responded to his pressing, with its barely-concealed challenge. But he seemed too honest and straightforward a man to be familiar with the deadpan ego games with which some Western climbers indulge themselves. I would be making for the Gangotri Glacier, but I felt it was essential to move to my own rhythm, not someone else's. It pays.

The Bengali realised that my mind was as made up as his, and rose nimbly from the table. "Then we go," he announced in a firm voice, giving his porter a few words of instruction in Hindi. "I have to meet elderly man…show him round," he added by way of further explanation.

He seemed not in the least offended by my refusal to continue. Thus I restrained myself from reminding him that while I had just walked thirty kilometres, he had covered the same stretch of road by bus!

We shook hands and parted company.

My way continued down to the wood and metal bridge, on the far side of which rose tier upon tier of huge, forest-hung cliffs, some of them stained by black drainage streaks hundreds of feet in length. From the topmost tent-like ridge, a translucent orange vapour streamed into the high thin air like some medieval banner of war – would the weather hold? It was difficult to say, although one thing was clear to me: the early evening temperature was far lower than I had encountered at Uttarkashi. But this was to be expected since there was a difference in altitude of some six thousand feet between the two places. I could only hope that my minimal warm clothing would be up to the task.

A second and smaller, bridge spanning a narrow ravine rapidly brought me to a stone enclosure and a neat, handwritten sign in red: YOGA NIKATAN. I had first gazed over this wall in 1991. This time somebody else appeared to be in charge.

Sitting before the open doorway of an Alpine-type hut was a grey-haired swami in white robes. He was engrossed in a newspaper and didn't notice me until I pushed open the gate in the wall. A high-cheekboned gaze lifted enquiringly in my direction as I entered the enclosure, relieved to be within sight of rest at last. He must have sensed my end-of-day state, for he addressed me simply, thus:

"Welcome. Can I get you an apple?" A smile accompanied his question, putting me immediately at ease.

I couldn't think of any good reason to refuse his offer and smiled my thanks as he rose to his feet, motioning me to the unoccupied chair in front of him. Perhaps 60 years old, he was of imposing height, and disappeared inside his hut with fluid, unhurried strides.

The apple, when it came, was about the size of a small pumpkin, and quite the largest I had seen since leaving the Western hemisphere. I had my suspicions that the swami thought I looked undernourished, but if that was the case he was too polite a man to say so. The fruit was big enough to make a pie on its own and presented a daunting prospect if custom dictated I should eat it all on the spot. I gave it a determined-explorer look before taking a jaw-straining bite. A second surprise. In Rishikesh I had avoided buying this variety assuming from its colour that it would have the teeth-clenching sourness of a cooking apple. Quite the opposite was the case: it was by fare the sweetest I have tasted in a long, long time.

Noticing my sudden change of expression, the swami commented, "Yes, they are very good apples here." He disappeared again, to re-emerge carrying two glasses of chai. Grateful for the timely opportunity, I planted the fruit-cannonball on the top of my backpack and thanked him for his kindness. He must have guessed I was tired from my journey and spared me the tedium of the usual cross-questioning, restricting himself to asking where I was going.

"You are going to Gaumukh tomorrow? Why so much hurry?" He glanced up with a clear gaze at the empty sky above the garden. "Tomorrow the weather will not be good. Stay here and rest. The day after will be fine – you can go up then."

I must have appeared unconvinced, for he added, "If you get to Bhojbasa, the ashram before Gaumukh, what will you do there? You will not be able to see anything – it will be worse than here!"

If he was right about the weather, then his recommendation made sense. Nevertheless, I decided to postpone my final decision till next morning.

When our glasses were drained the swami rose to his feet:

"You can get some rest before evening meal – we ring gong six-thirty." He indicated the row of weather-bleached, A-frames twenty yards from his own. "You'll better take a key."

Apart from the absence of any kind of heating, the ten by nine chalet was a sufficient shelter from the fast-falling Himalayan night. Built from sturdy, double-walled planking to assist insulation, it contained two neat bunk beds and twin doors which opened inwards. In the back wall a small shuttered window overlooked a neighbouring garden, decorated with unusual sculptures fashioned from natural materials. The bed, though narrow and about as yielding as a kitchen table didn't prevent my dropping upon it and falling asleep for the best part of an hour.

On my previous unsuccessful attempt to reach the Ganga's source, I had unwisely accepted the offer of a smoke from a passing *chillom baba*. Mountain air is heady enough and the combined effect of rarified oxygen and strong local-grown *ganja* was almost instantaneous: the pot I was stirring, the chalet-kitchen, the trees, mountains and sky, the voices of other visitors – all began to swirl together in a mad vortex of dizzying speed. My head pulsed with spinning lights as if I was part of some mad fairground ride which might at any moment spiral off into the black unknown: my forehead turned cold and wet and I could no longer feel the spoon I was holding. A dim feeling, a tiny ray of light penetrating my consciousness, told me <u>I must lie down</u>. All else was consumed by that single, urgent mission, to lie down.

The author's accommodation, Gangotri, 10,000 feet.

The spectacular approach to Gangotri.

Three hours passed.

Finally, when my body at last stopped sweating, I raised myself upright and quickly scribbled down what seemed to be an inspired opening to a short story. Not until next morning did I grasp how easily I had deceived myself. Later, when I complained of diarrhea to the swami, he gave me a foil-wrapped pill which he said was "for long bus journeys". Only when we encountered pot-holes on the road an hour or two later did I discover that he had given me a laxative.

I was under the impression that I was the sole occupant of the dozen or so chalets this time, but soon after the sharp tones of the meal gong faded, I heard footfalls emerging from nearby rooms as shadowy figures converged upon the rustic table directly across from the kitchen.

Four nationalities wrapped in an assortment of winter clothing were represented at the evening meal: a middle-aged German man from Munich, an English woman in her seventies, a stubble-faced Italian with a mild form of facial neuralgia and a young Malaysian who answered every attempt at conversation with a polite smile. For the first time I was fully clad in my 'high altitude' outfit comprising a pair of gloves, lightweight long-johns under my Delhi cottons and an anorak covering a quilted waistcoat. In reserve in case the weather turned rugged later on I had a scarf and a pair of snow goggles.

Perhaps the near-zero temperature was the cause of it, but a decidedly formal mood hung over the outdoor table that evening, the wan proceedings illuminated by the pale glow from a paraffin lantern supplied by the hospitable swami. Mountains attract the hardy and self-sufficient – often the emotionally self-sufficient as well, and little if anything was said while we waited to be served. I had only taken a snack at the bazaar across the river and was grateful for the hot food when it arrived. My strongest memory of the event was a visual one, a row of silent, deeply self-absorbed figures – and the alert swami striding back and forth with offers of second helpings. Beyond a chilled nose here and an uncovered hand clutching a spoon there – was the darkness of the encroaching forest. Occasionally a faint breeze would stir, spreading the penetrating fragrance of pines amongst us. The Malaysian was the first to leave, dispatching his food without pause, before swiftly disappearing into the darkness with the briefest of goodnights.

I was wrong in thinking the German was travelling with the elderly English woman. After a protracted pause to finish his tea, he rose spectrally at the darkest end of the table, saying, "I go talk wid dos *baba* men by der river. Here it is ver quiet; nothing is happening."

What exactly he wanted to happen, he didn't specify. I couldn't see any sign of a beer keller.

Myself and the English woman were the last to leave. She told me she had moved to Andorra years ago to escape the British weather and liked to spend three months annually in the Himalayas.

As I was rinsing my plate under the tap by the kitchen veranda, the swami cautioned me about taking nocturnal strolls beyond the ashram gate: "-Part of the footpath collapsed in a storm; now there is eighty-foot overhang – very dangerous. In summer American woman walking there, taking photographs. Fell over. Broke all four limbs."

Cheerful news – but he needn't have worried. I had done enough walking for one day, my arms were cold and, despite the relatively early hour, I was ready to dive into my sleeping bag.

I took a final look at the night sky from the threshold of my cabin door. No moon shone as yet in that glittering, overpowering dome of stillness. Not a cloud was to be seen – even the few shreds I had spotted earlier flying from the highest ridge had vanished. But the sky! It sparkled with more stars than I could remember, so pure and clear was the atmosphere, while in the deepening cold, the nearest peaks had begun to glimmer like the phosphorescent walls of some visionary palace. It was another of those all-too-short experiences when I felt the earth's energy, her aura, resonating in harmony with my own. The swami's forecast that the morning would bring deteriorating weather seemed unnecessarily pessimistic: to me the air temperature felt far too low for snow.

When I awoke next morning from a strangely-muffled dream of home, I saw promising slivers of brilliant light piercing the cracks in my cabin door.

I could not have been more wrong. As soon as I opened the window-shutter above my bed I was greeted by a transformed world which scarcely resembled the one of the previous evening. The walls, the fruit trees of the adjacent garden, and the forest – all were coated with a thick dusting of fresh snow. It was my turn to bow! A few degrees rise in temperature had brought with it an all-obscuring mantle of snow-bearing cloud.

A sinking feeling filled me as I forced my feet into my freezing boots. The Himalayas had once more revealed its other, more hostile face – was it going to scuttle my second attempt to reach Gaumukh, the 'Cow's Mouth'?

In my dream I was travelling alone on a slowly-moving train through the heart of an industrial northern city. It was the 19th century. Rising above the street running parallel to the railway line were the shapes of buildings as though seen through a veil or morning mist. A tram passed containing passengers dressed according to Victorian times. A man on a very old bicycle slowly overtook the train, staring directly ahead as if focused upon his destination. I felt surprised that none of the other passengers seemed to notice him. At that point the dream began to fade and the rider gradually melted away into the deepening mist.

At breakfast I was heartened to notice the swami remained unperturbed that visibility was down to treetop level:

"Tomorrow there will be no problem," he assured me, as he calmly served bowls of porridge. "If you go up today the path will be slippery with wet snow; there is a risk of avalanche and every year people are killed. There is no harm in staying a day or two in Gangotri, though these days few people do so. They prefer to rush straight up to Gaumukh and back again, just to say they have been there."

I decided the swami knew his stuff and asked about the deep ravine, a vertical gash which split the mountainside above his ashram. It looked to me like an interesting objective for a day's local exploration.

"It is the trail to Kedar Tal, a beautiful lake at 5,000 metres, but I wouldn't advise going without a guide. Unless you know the way it is very easy to lose the path, particularly in the lower part where the ground is very steep."

He spoke in an unsensational way, as though I could take his advice or leave it.

"Only last week three Indian men went up there in the late afternoon. One of them slipped and fell seventy feet on to a ledge. He could move neither up nor down; his friends were unable to reach him. They threw down all their warm clothing to this man and came here to fetch help. But it wasn't until next morning that a rescue party was able to bring him down. Now that man is in Uttarkashi hospital more dead than alive."

After a sobering interval he continued in a more cheerful vein. "There is a cave which you would find of great interest. Follow the river downstream for forty minutes and you will come to the next valley. It is there. That place is also the way down from Kedar Tal."

It <u>did</u> sound interesting but my ears weren't the only ones to have pricked up: the Italian trekker was leaning in, an eager expression flickering across his twitching face.

Alive to likely developments, I made only a non-committal response to the swami's helpful proposal. And sure enough, as soon as the monk turned his attention to other matters, the Italian addressed me in strange, almost inaudible broken English. He asked if I was thinking of going for a walk, implying of course, that we could find the cave together.

It was an awkward moment, I sensed he didn't really care to be travelling alone and was on the lookout for company – anybody's. I didn't want to hurt his feelings more than I had to, but I trusted the voice of my instinct which said I would be better to visit the cave alone. I didn't want to ask him along for politeness's sake, only to find I was carrying the extra psychological burden of a companion who needed me more than the walk.

I changed the subject after putting him off, asking the swami about my earlier visit when I recalled another monk in charge of the ashram.

His face lit up. "Ha – you mean Swami Ram Brahmachidanya – he is still here!"

"Here, in this ashram!?" I glanced about.

The swami shook his head, his gaze veering towards the gate, beyond which we could hear the rush of the river through the narrow gorge. "He _was_ the manager of this ashram. But now he is in charge of Yoga Ashram" He made a diving gesture with his hand – "If you look down from path when the sun is out later, you will see him in his garden. You can speak with him when you return from cave."

I set out from my room shortly before nine o'clock with my enthusiasm for visiting the cave undampened by the conditions. I wasn't expecting to be away long and carried no food apart from an as yet untried _prana_ ball, a few dried dates and a little loose tea and powdered milk

There was no obvious footpath as such, only an irregular sequence of clearings amongst the trees which somehow linked up. The mist-shrouded conifers were of no great height, but they soon screened the ashram and other human habitations from view as though they had never existed. Other than my own size nine's I saw no footprints on the intermittent patches of damp snow. The forest was as deathly silent as the river was noisy, but so long as I did not stray too far from the latter, I knew I could not go far wrong.

After a while I noticed the tell-tale signs of past human occupancy beneath a granite outcrop, and in another instance, amongst a clutch of large boulders of this kind of rock. But they were not what I was searching for and I gave them only a superficial inspection. Snowflakes continued to drift gently out of the luminous grey void which hid the high peaks from view. Winter had not yet officially begun, though, and most of the flakes melted the instant they touched the ground.

At last I spotted the humped form of a rock far larger than any I had so far passed. It was a towering boulder weighing many hundreds of tons; its top, curving swiftly to perpendicular flanks reminded me somewhat of the blunt head of a sperm whale. A weather-tattered flag denoting the spot to be a place of spiritual sanctity, hung from a bleached staff, planted at the rock's highest point.

I approached closer and saw at the foot of its nearest face, a small wooden door partly protected by a natural bank and the overhanging nature of the speckled rock. A few yards from it, a rough circle of irregular blackened stones marked the remains of a long dead fire. That this was the place I had no doubt.

Listening at the door, I quickly came to the conclusion that no-one was within, and quietly raised the simple latch. It swung easily inwards, revealing the interior to my curious gaze.

Once my eyes became accustomed to the gloom I saw that I was at the entrance to a single large chamber. It is easily described.

The cave, formed by the smooth underside of the stone resting upon the ground at its tilted extremities, was about forty feet across in its longest dimension and as high as my chin in its middle portion. Clearly some effort had been made in the past to turn it into a dwelling, but there was no obvious sign that it had been occupied in recent days. On the left a platform of stones was raised against the uphill wall, most probably to serve as a sitting and sleeping space, while the uneven floor itself, sloped away at an angle until it met another door thirty feet distant, and even lower that the one I had just entered. With both of them closed, only a little light seeped into the cave.

In the mundane sense of the word the cave contained nothing. A casual visitor would most likely take a swift glance around before ducking out again to dismiss it as 'empty'. It is true – there were no polished walls as at Marabar to set your lamplight dancing as in a hall of mirrors; no artefacts or personal items discarded by previous inhabitants to wonder at. Examine the walls for as long as you like and you will not find a single line of scriptural inspirations written there, no graffiti, no drawings – nothing. If you raise your voice, making a high, musical note, no nine-second echo spiralling into infinity responds, as it would in the dome of the Taj. In fact there is no echo at all. Particularly if you go there in company, you might marvel at the size of the cyclopean boulder and its situation, but find the cave beneath almost dull.

Yet it has something. If you sit inside for a short while, remain still – you begin to feel its presence. Within one's own ears, perhaps a ringing sensation or a rushing noise begins to arise. You might think you are simply hearing the sound of your own blood circulation or that your family history of tinnitus has finally caught up with you. But if you listen steadily, just letting it happen – other sounds more refined than the first begin to come. Maybe you will hear bells or the rarified tone of some highly-tuned wind instrument…One's body relaxes as the mood begins to stabilize; stray, spontaneous thoughts become less frequent until finally they vanish altogether. Only when your thoughts fade do you truly enter the cave and begin to experience its sheltering, transcendent spirit. No, the cave is far from 'empty' in that sense.

There are other caves, perhaps millions of them, which have never been visited. Locked deep within the Earth's crust, they have never seen the flame of a torch or felt the wind's breath caress the walls of their labyrinthine passages. Now bats live in them. Inscriptions do not disfigure their unplumbed interiors because no-one has been there. Not even those who dare to descend the deepest cave in the world have any inkling of their presence. These caves were born far back when the great chain of the Himalayas was still upthrusting, and Mother Ganga had not yet let down her tresses. Aeons before the wheel caves were. They are as dark and unknown as another universe. Now and again miners, burrowing like termites into the Earth's skin break unexpectedly into one of these chambers – caves which no human eye has ever seen. They can be as small as foxholes or larger than several rooms stacked one upon the other.

A few years ago I myself had a strange experience in one of these caverns or 'vughs' as the miners call them. Whilst exploring a maze of 19th century mine workings amongst the treeless hills of northern England I came across the cavity in a disued tunnel or 'level' which had been driven along the line of the vein of lead ore. Torpedo-shaped and about twenty feet in length, it was high enough to crawl into through a hole at one end. In the beam of my lamp, the inner walls glittered with crystals of various minerals including coloured fluorspar and quartz.

Just as I began to inspect these unblemished treasures which clustered like grapes all about me – my lamp flickered and went out!

It is easy to feel a creeping panic when darkness and silence begin to merge into one: unless there is running water, the silence in old mines is absolute. And the darkness…blacker than soot and heavier than air, its density packs suffocatingly close, a palpable thing which quickly blinds rational thought. My fingers closed about the squat button of my caplamp…squeezed, and turned with a prayer. Utter relief as the unimaginable blackness sprang back into the walls like a nightmare spider withdrawing its legs.

Many who have wished to attain their highest spiritual goals have practised their transforming exercises in the purifying surroundings of mountains, forests – and caves. In the 'developed' world the idea of dropping everything in order to seek some higher 'reality' in such surroundings, is tantamount to the first signs of mental derangement. But in the forest near Rishikesh I met a German woman who had made that choice.

Originally an artist from Berlin, she had lived in a cave high above Laxman Jhula for seventeen years with only two huge Tibetan dogs and occasional visits from an Indian helper for company. Her reputation as a woman of great spiritual attainment was widely known amongst Hindus. Many visited her – or tried to, for her secluded retreat was intended not to be found easily. Those who succeeded treated her with due respect, but gradually the solitude which had first drawn her to that place became more and more difficult to maintain.

I discovered her sanctuary at a bouldery crossing of a stream, with tangled tree cover on the neighbouring slopes. I could see the cave-dwelling twenty yards above me, but not wishing to arrive in abrupt fashion, I made a prudent halt. "Why hurry," I thought, "when a woman has already spent almost twenty years of her life in this one place?"

Even as I loitered by the stream, however, the subject of my visit emerged from a door, a small, energetic-looking woman perhaps in her late forties or early fifties. She spotted me almost at once and invited me to join her on the terrace. Did I want tea? I was pleasantly surprised to be made so welcome after what I had heard. Nevertheless, it quickly became obvious that the German was not a gregarious sort, and it would be wise not to overstay my welcome.

The cave, guarded by a normal-sized door, stood at the right hand end of the narrow terrace; a stone extension with a flat-roofed upper-storey had been added at some stage, a construction of minimal refinements, but entirely fitting for what it was – a hermitage. As I crossed the threshold a warning growl caused me to look up, straight into the glowering gaze of two, bull-like heads framed in a square-cut hatch. The German uttered a short command in Hindi, whereupon the dogs lowered their huge heads gently upon the rim of the aperture, while their eyes continued to burn in my direction with unwavering hostility. When I queried the need for such brutish-looking beasts, she replied:

"It's safe enough here in the day, but at night panthers and other animals sometimes come. Also there are people who would like to rob you."

We drank the tea sitting cross-legged on lengths of folded sackcloth outside on the sun-dappled ledge, now and then exchanging a few observations concerning the value of the solitary life. I rapidly gained the impression that here was no complicated individual living an entangled life in simple surroundings, but a woman of enviable integrity and intelligence, living the spiritual life, hour by hour, day by day, year after year. Her spasmodic talk did not float off into the stratosphere of idealistic abstractions, but remained lightly rooted in the present, the here and now. Some Westerners, ever-conscious of the impression they are making upon others in India, throw a cloak of spiritual eccentricity about themselves, but I could see that this woman's strength was an inner quality; she had no need for outward display.

Inner development is more than a system of esoteric exercises marooned in an ascetic life of self-denial. If it means anything at all, its presence reveals itself in unobtrusive details – a glance, the movement of a hand or in the tone of voice. I could easily sense why she had become an inspiration to so many.

For the first few years, she told me, she would return briefly to her home country every twelve months in order to make enough money from her art to meet her basic living expenses for the following year in India. Today, the generosity of well-wishers meant she had not needed to make this long journey for twelve years. Nevertheless, there was a price to be paid for becoming a kind of 'spiritual celebrity'. Even as we sat in the silent forest, men's voices and once a woman's, called enquiringly from far down amongst the trees.

The German looked towards the sound: "They are trying to find me," she smiled faintly.

We listened in silence as the calls drifted back and forth, now louder, now fainter, before finally fading slowly into the distance as a leaf is carried away on a stream. My attention switched for a moment to a tiny bird, a dipper, hopping amongst the polished stream rocks.

The woman spoke again. "I shall have to move soon."

"Why is that – the people?"

"Not only. The regional government has decided to turn this area into a state park – I will not be allowed to live here any more."

I detected only a hint of regret.

"In a few months a friend will come and I will be gone from this place. I have already found another cave less than forty kilometres from here. I will move there."

Her room-sized cave was a kind of limestone grotto, generously decorated with stalactitic drapery overhead, whilst standing on ledges were several curious stones, each resembling a living entity and formed from the infinitely slow drip of carbonate minerals. In order to give the door a more secure attachment, cemented stonework had been added to the left of the natural opening, above which was a canopy of overhanging rock. Sporting a playful attitude in an adjacent niche was a human-sized cement sculpture of an androgynous Hindu deity, probably *Siva* to judge by the 'dancing' display of its four limbs. A companionable dog with head raised crouched close by, likewise fashioned from cement. The potential of some of those weird beings to arouse fear was balanced by the colourful presence of more familiar temple decoration, along with simple votive offerings daubed in vermilion. In Buddhist symbology, the 'demonic' resemblances would be regarded as those aspects of our minds which have to be faced up to, before Emptiness, *Nirvana*, can be realised. A few domestic items such as cushions and rugs added a touch of softness to the interior; it was not simply a chamber of idle curiosities collected by an artist, but a space of true spiritual devotion.

When I departed after a two-hour visit the German called after me:

"Next time you can bring me some fruit!"

Perhaps we are meant to cross tracks with such souls only once. They act as truth mirrors to ourselves. In them we see how far we have come…how far there is still to go. Sometimes upon the trail or in my sleeping bag at night, memory calls them back; and in the flickering light of my inner eye I would see the slow procession of their half-remembered faces once again.

The Gangotri cave was different altogether. For some time I sat without moving and with my eyes closed. Whereas at Laxman Jhula the German's cave had pulsed with the devotional ardour of its occupant, the opposite was true of the cavern I was presently visiting Of the imprint of any human personality or sign of a lonely individual's inner struggle – there was scarcely a vestige. Impersonal, unattainable to mortals – whatever the vibration, this particular cave simply <u>was</u>.

Alexandra David-Neel tells many stories of mystic-hermits who lived alone in the wilds of central and northern Asia. She writes of those who attain magical powers through prolonged periods of confinement, often in walled-up caves. By means of 'one-pointedness' of attention the mind is freed from its attachment to many kinds of ephemeral objects, including passing thoughts and feelings. By eliminating both internal and external distractions, consciousness itself becomes focused, and heightened states of awareness – beyond mind itself – are attained. The initiate sitting in his sunless cell suddenly finds his room is filled with a sourceless light. Ordinary objects in the darkened room begin to shimmer as if illuminated from within. Everything is in vibration, nothing is solid. He feels he is part of the living stream. Psychic powers may arise during such states.

David-Neel writes mainly of Tibet and it is not within the compass of this book to describe in detail the training which Tibetan mystics undergo to develop their extraordinary powers. But certain aspects are clear and such knowledge is of course not the exclusive property of Tibetan culture.

Apart from the presence of ghostly conifers, there was not a lot to see in the immediate vicinity of the cave. But it was too early to head back to the ashram, so I decided I would first make a rapid reconnaissance of the narrow ravine upon whose edge the giant boulder so neatly perched. In my favour was the fact that it had stopped snowing and the mist appeared to be thinning. Soon I reached a tree-fringed notch which formed a break on the edge of the ravine. I could hear the sound of pounding water below and peered down: two hundred feet. It looked possible with care, a ramp or rock ledge sloping from the gap suggesting a good place to begin my descent. My own survival instinct is as strong as most people's and I proceeded with due caution along the edge.

I soon realised that my 'path' was actually a natural weakness marking the junction of a double-decker cliff. It was a promising start, but upon rounding a projecting cheek of rock after only fifty feet, I was dismayed to find my ledge terminate in a corner, with a greasy, fifteen-foot slab immediately below. At the same moment I spotted evidence that others had used the way before me: three or four aged poles had been flung against the corner to form a crude ladder. It looked to be on the point of collapse, but there was no other way. After a moment's hesitation to assess the landing, I lowered myself from the ledge and slithered down the poles, giving my trousers some interesting stains in the process.

My landing, though safe, was now even more sensational. Black, glistening slabs, thinly vegetated, fell away beneath me, ending at an abrupt void in whose depths boiled a thundering cauldron of spray. It was an exposed situation but I had already gone too far to think seriously of retreating the same way. Over to my right as I faced outwards, the angle though still slippery was less intimidating, with a few stunted bushes providing encouragement for use as emergency handholds should I need them. The bushes, when I reached them, proved to be liberally sprinkled with gooseberries in a perfect state for eating, a little more 'tart' in flavour that I was used to, but I couldn't complain in the circumstances and devoured the unexpected treat with relish. Lower down I also chanced upon an unusual wild strawberry plant with a dozen or so crimson fruits clinging to each upright stem. Like its neighbours, it had received the attentions of some creatures or other, possibly the goats which roam wild in this region.

At the bottom of the ravine I was met by a jumble of glassy boulders, spray, and a cacophony of sounds reverberating from the imposing scalloped wall which faced me across the torrent. I also discovered something else invisible from above – a bridge. Four, twenty-foot tree-trunks spanned the stream, forming a level-enough platform for me to cross, any gaps being plugged by wads of packed earth and stones. On the far side a steep rise led me beneath the weighty limbs of a towering cedar to level ground, and once more within sight of the rushing Ganga. Goat droppings and the remains of a couple of fires told me it was in use by the hill shepherds and their flocks. I cast around for awhile above the camp until I found what I was looking for – a trail leading up through the trees overlooking the right-hand cliffs of the gorge. It could only be the way down from Kedar Tal, which the swami had been telling me about.Now it was time to return.

Wandering back the way I had come I had two unexpected meetings, the first being a sight of a large Himalayan hill fox, bigger than any I have seen back home.

As well as those which inhabit the foothills such as the tiger, the elephant and the quietest of them all, the panther, there are also the black and brown bears, the leopard, the goat and the ibex, to list only a few. Perhaps the most intriguing is that much larger animal of fable and myth – the yeti, sasquatch or abominable snowman. Numerous sightings have been recorded over the years, but to date there is insufficient evidence to satisfy a sceptical age in which 'seeing, weighing and measuring is believing'. Until one is captured and either put in a zoo or appears before a scientific board of enquiry, the situation is likely to remain unchanged.

An American Buddhist monk who had lived several winters at a monastery in the Mount Everest region told me:

"Yetis do exist. Sometimes when our monks make solitary retreats high in the snows, these creatures come near. They are very intelligent and are drawn to the monks, sensing their harmlessness and purity of purpose. One of our senior lamas was meditating alone in a cave for several days. He became aware of a presence near the cave mouth. Whenever he went outside, the creature had vanished; but there were footprints in the snow. If he moved to a different cave he was aware of being shadowed from a distance. His impression was that the creature, shy though it was, was acting as a kind of 'guardian'."

The American had not had a similar experience himself, he told me, but in the long winter nights he had heard their cries outside, from the monastery. "If you try to find them as part of a TV team with your cameras, noise and cans of beer – they sense telepathically where you're at, and stay away."

In a brief correspondence on the subject, Sir Chris Bonington, Britain's best-known mountaineer, who himself returned from such an expedition without tangible proof, wrote:

"I am quite sure that the yeti can do without ardent journalists and television teams. I'm quite confident he will sense the intentions of our busy media men, armed with their image intensifiers and trip wires and keep well away. I suspect the best chance of seeing or meeting a yeti is to go up into the high mountains and just sit there and project good thoughts, and certainly not carry a camera."

Perhaps they are more likely to find us than the other way about.

My second encounter was with one of my own kind, the eavesdropping Italian. Footpaths mean other people and my spirits dropped when I saw the hunched figure making his way slowly towards me through the trees. His cheek twitched as he came up to me with a melancholy look. Whether he had a speech defect or was simply inhibited by a lack of common language, it was difficult to tell. He seemed to be filled with a kind of internal pressure, as if he was bursting to share something of extreme importance, but didn't know how.

A spasm shot through his jaw as he struggled to speak. At last a single word, barely audible, fell from his lips:

"Walking…" He looked at me hopefully.

I stared back, unsure whether he was asking a question or merely affirming what we were both doing: walking. Did he wish to walk with me, or what? I shifted and looked beyond him.

The Italian's colourless mouth was moving again. His eyes flickered past me, a wan smile hovering uncertainly in his face, as if he was afraid of something. "C..cave?" he queried doubtfully.

"Si, cave," I nodded, relaxing. He was wanting us to be trekking buddies, was my impression. I stabbed a forefinger downhill towards the narrow footbridge:

"Me bazaar! *Khana* – samosa. Si?"

His face widened with the dawn of understanding. "Ah, samosa!? *Avanti* – not cave!?"

"Si, avanti."

For a second I thought he was about to join me; the situation was faintly absurd.

"Cave good?" he laboured as if cold.

"Si – good!"

"Uno?"

"One hour, si."

"Ashram sleeping?"

"Yeah, ashram."

"Ahhh – Gaumukh?"

I shrugged and didn't reply. Normally I feel late thirty something but in the Italian's presence I could feel the years creeping up on me. With a great effort, I tore myself away.

"You'll find it, all right!" I called back as I made off, a shade more brusquely than was necessary. He needed a companion too much, the Italian. Or a counselling session.

Five minutes later I halted at the middle of the bridge and leant reflectively over the railing. It's strange, just as you start telling yourself you're THERE, things happen. Anyone can lie in bed thinking beautiful, kindly thoughts about all living beings, sentient or otherwise. And how much compassion do you need to sit meditating, when all the others in the circle are breathing slow and deep, just like yourself, as if everyone was already a self-realised spiritual master, rock-steady in all situations? When you least expect it, the door flies open and in lurches the Italian with his irritating remarks and questions to test you out. How patient are you then? Is your compassion only for those who don't say stupid things? Perhaps such people are our measuring sticks.

After leaving the Italian to his own devices, I decided the Himalayas needed cheering up at that moment: it was time to treat myself to one of the *prana* balls mentioned in an earlier chapter. But even as I peeled back the first layer of metal foil wrapping, I knew I was again out of luck as a faint odour pricked my nostrils.

I stared for a few seconds in disbelief – my mega-cookie was covered in greenish mould like a ball of rotten cheese. Shit! I swore again before tossing it over the railing, watching it pancake against a rock twenty feet below.

In the middle of the afternoon the clouds drew swiftly aside and the wonderful Gangotri valley was flooded by warm sunshine. Not only that, but just as the swami had predicted earlier, Swami Brahmachidanya was sitting outside in his sunny garden forty feet below the footpath. He glanced enquiringly from his books; his long black hair streaked with silver was tied back in a bunch of dreadlocks.

214

"We spoke together five years ago," I called. "Can I come down?"

Recognition suddenly filled the swami's face:

"Yes, I know you! We met on the road – you had just got off the bus, I remember the incident well!"

He beckoned enthusiastically, "Yes, yes! Come down, come down!"

It was a real joy to be made so welcome for a second time. After a short exchange of news we began to talk as though 1991 was yesterday.

"Last time you told me that Indian people have lost their way, spiritually?"

The swami nodded tolerantly.

"The majority of people come here in groups. They come just to see - "

"Curiosity?"

"That is correct. They walk about, take photographs – they go home. They have seen it. You ask them about the Vedas: they will know that there are four books. They do not read them. My friend, they probably know less about the Vedas, than you do."

He spoke as one who had been through much himself, both in the world and out of it. His former profession as an engineer lasted until his fortieth birthday, when he travelled to Haridwar, threw all his money in the Ganga and walked to Gangotri to take up a new life as a monk.

The swami gestured at the scriptural text which he had laid aside.

"It is very difficult for all of us – it is difficult for me, too... It is the world today. Even for me it is not easy to remain steady. I am not an agitator – I do not go out on to the streets to demonstrate. My job is to tell people who, like yourself, come to visit."

He spoke about how he had expended much energy over a number of years in trying to raise the consciousness of people in power, to make them aware of the effects of tourism on places like Gangotri.

"Since the new bridge was built more and more people come here. Before, only those people whose primary purpose was pilgrimage, hardy persons, would make the long journey to Gangotri. Now we get tourists. They also pay their respects at the temple. But pilgrimage is not their primary purpose. In the season they come by taxi, bus and jeep…there is no peace any more. In the winter when there are only three or four of us living here, I like it much more."

The swami regarded me for a moment.

"I told the local governor what is happening up here – 'the solution is in your hands'. I told the local people the same. There are brown patches – pollution, high on the mountains, when before, the snow used to be white. Stop the traffic at Harsil. Stop all the buses and cars and Gangotri can still become what it was – a nature paradise."

He smiled ruefully.

"Everyone said, 'What a good idea; stop all the traffic'. But then there are some very poor people who depend on these tourists. Each season when the temple opens in May, they come up here and hope to make a little money. Who am I to tell such people what they should or should not do?"

"Perhaps you have sown the seeds?" I suggested.

There was a pause.

"I have thrown the seed. Whether it will land on fertile ground, who can know?"

I referred to the Tehri dam. "Can it still be stopped?"

"How can it be stopped when there is so much money involved? People – they do not realise that the money is all 'Dead Sea Apples'."

I looked puzzled.

"You know 'Dead Sea Apples'? Fruits which are clean on the outside, but dead in the middle – dirty money."

The swami was translating scriptures every day and I hoped I wasn't keeping him from his work. I glanced around his orderly garden, wondering if it was all right to stay a little longer. Here and there, a piece of branch with unusual grain patterns or animal features had been carefully placed to catch the eye. In a stone, I saw the striking profile of a human face. No sound came from the four or five cabins which seemed to comprise his ashram.

"So you are back!?" the swami laughed, interrupting my thoughts. "You go to Gaumukh?"

"Tomorrow," I nodded. "Tapovan, also – I hear it is a very special place. I go alone, otherwise you don't feel it."

He agreed:

"You are right to go alone. Or at most with one other person who is there for the same reason."

"It's funny," I said, "there's so many places to go in India, but I come back to Gangotri!"

His eyes searched me in a kindly way.

"That is because there are lessons you have to learn here. Everything that is going to happen to you is already mapped out. That is why you are sitting here in Gangotri, at this moment, talking to me."

"I almost left this morning," I interrupted.

"But you stayed?" came the calm reply.

I looked at him for a long time.

"Whether you think that you personally are deciding this and that, the game is already laid out for you to have certain experiences, in order that you ultimately come to understand who you are." He raised his eyes skyward. "I don't mean there is a God up there with four arms, saying, 'DO THIS! DO THAT'! No; it is the Cosmic Plan ordained by Nature that in the end we come to realise we are all part of the Greater Whole."

I didn't have any problem with what he was saying.

He spoke again.

"There is the river flowing by. Take a bottle and fill it from the river. Can you now tell me that the water in the bottle is not the river, one and the same? You can try." He continued without waiting for a reply:

"But then if you break the bottle."

"Like the wave which has separated itself in the ocean?"

"Exactly. It merges back with the sea, becomes part of the Greater Whole. But it will arise again and again, giving a semblance of birth and death."

"Many people in the West – Christians – have difficulty with *Karma*. Or they use it as a stick to beat someone with. I'm not sure I completely –"

The swami smiled. "Yes, many Westerners want to ask me about *Karma*. People are always saying, 'Yes Swamiji, but what about all these famines and floods – what did the people do to deserve them? What if there is a bereavement in the family'? No, of course not! At the time of personal bereavement you cannot just tell that person it is their *karma*; you cannot simply tell the family that the person who died – perhaps very young – did so as a consequence of some past action. That would be wrong. They are so involved with the situation, in their own suffering, that they would not be able to receive it. Better to cry with them for their loss."

"Then later?"

"Later on, they may be ready to accept." He raised one finger. "But it is not fate; *Karma* is not fate, you should be clear about that!"

There was a pause as he changed tack:

"The most important question you can ask yourself is, 'Who am I'? What is your second name?"

I told him.

"'Clark', that is your clan; you belong to that clan. Now. Remove 'Clark' – immediately you get rid of the clan. What is left when we drop our names, positions in life, and so on? Until we stop saying, 'I am this', 'I am that', there is still the ego. Only when the 'I' finally disappears, do we have pure Being…a continuous present. We live in the moment: past, present and future are all One."

"Existential?"

"Pure Being. No Mind. PEACE. BLISS. JOY. Have you experienced – ?"

I mentioned meditation, music. "Music isn't just the sounds but the spaces between the notes. Sometimes in meditation there are these gaps also, when the thoughts don't come."

Leaning forward, the swami nodded approvingly.

"In mediation, the one who experiences Bliss, he sometimes stops breathing altogether. He is totally removed."

"That's the dilemma, isn't it?" I responded, " being in that state, yet continuing to communicate with the world?"

"The man who is in this state of consciousness, *Samadhi* – he cannot. It is impossible!" He paused to allow his words to sink home. "What normally happens with such people is that they go into *Samadhi*; then they come out of it!"

It sounded perfectly simple. Swami seemed to read my thoughts. His eyes narrowed until he was almost squinting. For a split second I saw myself sitting opposite a Tibetan shaman from another age. His eyes drilled into me.

"This is not just an idea, one of your Western hypotheses; it is fact!"

A strange sensation filled my head. I saw spiralling forms amongst the nearby twigs and branches which hadn't been there a moment before. An accidental arrangement – ? Was Swami merely trying to convince me? No, he didn't seem the kind of man to acquire disciples. There was more behind his rationality, much more...

"And these moments of 'No Mind' are when powers can arise?"

Brahmachidanya's dreadlocks jostled in lively fashion.

"I ignore them completely. Don't try to get powers! They will come anyway. But if you use them in the world, they will go."

He mentioned the importance of breath control as a means to mastering terrestrial life and cited the examples of hibernating animals which sleep for months, a sleep in which breathing is suspended. If animals can do so, why not humans? I thought of *fakirs* being buried alive. After many days of guarded burial with all bodily orifices sealed, they would be exhumed, and in a short while be living again normally, as though nothing had happened. The acquisition of such control is boggling for the Western mind to contemplate; the perceived truism that breath is a condition of life is a notion not easily abandoned, no matter how reputable the Master who tells us otherwise.

A pregnant pause followed in which I sensed Swami did not wish to talk further about mystical powers and how to attain them, fascinating though the subject was. The astonishing field of esoteric knowledge around which our conversation had begun to turn, was not a topic upon which I felt bold enough to wax my own opinion one way or the other.

We both raised our eyes to the forest-fringed peaks towering overhead, and I once more began to hear the muffled rush of the Ganga as it plunged through the glistening ravine only yards from the secluded garden. Swami pressed his palms together and beamed a greeting as someone paused to salute us from the walled pathway above.

"A lot of people come here," he continued. "They want me to answer questions about the world. Even though I haven't travelled, seen other countries, they want to ask me."

"But you know the world, Swamiji," I interrupted quietly. "You don't travel, but –"

Chortling at my remark until crow's feet appeared at the corners of his eyes, he replied,

"Oh, I know the world, alright, I know the world and its ways! People come here…They say, 'Swamiji'," he mimicked a high, anxious voice, "'why don't you come down from Gangotri? No one hears you up here.'"

He laughed again, revealing a set of prominent teeth which reminded me of Dharmananda at Ved Nikatan. Westerners and their questions clearly amused him.

"People, they want to ask me about their own lives, about world problems. There was a Japanese woman last year. She said to me: 'You should come to Japan, my country! People want to hear you speak…Japan is very beautiful country; we could show you many interesting places: it will all be arranged'."

He fingered his plaited, two-foot rat's tail of a beard. His eyes brimmed with amusement:

"They look at this. They say, 'That is good, Swamiji – they will like that in my country'. I have been invited to Japan, to Germany, to New Zealand and other countries – they are only interested in my novelty."

"You would be on TV shows…driven around in a big limousine," I laughed.

The monk rolled his eyes as if for a moment the prospect appealed to him.

"Yes, I could do all of that!"

He became serious again.

"But I say to these people: 'If life will be so wonderful for me in your country, why do you spend so much money coming over here?' They look at me and say, 'Well, you see… Life in Japan, it is so very fast. There is no peace. It is more spiritual here…At home, no-one is very happy'."

I could see what was coming as he lit a short, black cigarette.

"Well, then, I tell them, 'If Gangotri is such a spiritual place – if I am happy here – why do you want me to leave and go to your country, where nobody is happy!?'"

I joined in his uproarious laughter. When it ended, he added wistfully,

"Yes, I know all these games…This is a peaceful place for me, Gangotri. I have found the work I have to do – why should I leave? If I go to Delhi, it is so polluted there, my health is bad. People can come here if they want to see me."

He looked at me keenly.

"One thing I have learnt: in every moment you live, never intentionally harm any living thing." He gestured around his neat enclosure. "That means I don't even move a stone from its place over there."

I agreed with a nod.

"One of the biggest dangers on the Path is spiritual hypocrisy," he continued. "Religious feeling without intellectual consideration leads to intolerance and dogmatism; we should always test our spiritual understanding with the enquiring rational mind – so far as thinking is able to do that."

I looked up at the sky. It was clear of clouds now. Hundreds of feet above, framed in the mouth of a cavernous ravine, two specks, eagles, circled and circled. With stationary wings outstretched they spiralled higher and higher on the up-draught of a thermal. Smaller and smaller they became as they rose higher even than the soaring orange walls. Our conversation was drawing to a close.

"What about mice," I asked, "do they have a place in our homes?"

Swamiji gestured to the open doorway to his right.

"I also have to deal with creatures inside. Mice and spiders – I catch them and throw them out." He shrugged as if the matter was of little concern. "Yes, they come back. If you kill them because they are making a noise, then you will have to live with it…your *karma*. If you then feel remorse, there is no need to go to a temple to tell what you have done. If you tell your own heart, things begin to change. We have to start with ourselves."

We talked for over two hours before I began to feel the need for a pullover on top of my T-shirt; the air temperature had dropped noticeably as the sun sank closer to the western mountain crests. I rose to go.

"Will you be here when I return from Gaumukh?"

Swamiji beamed as he said goodbye:

"I am always here."

A few minutes later a curious thing happened in my room. As I reached inside the open top of my rucksack for my jumper, something shot out and darted under my bed – it was a mouse. I could swear it was grinning.

Twilight was beginning to make phantoms out of the trees as I took to my sleeping bag for a brief rest before the evening meal. The best part of an hour passed before I stirred myself.

Swinging upright on the side of my bed I gazed steadily into the unwavering flame of my candle. With scarcely any effort on my part, I slipped readily into the state of *Tratak*, a powerful exercise in concentration known and practised by adepts for centuries. The technique not only strengthens the eyes, but has other benefits too. It can produce great calm and focus of attention as well as giving access to higher spiritual knowledge and control of subtle bodily energies (Prana Vidya). A place of peace and quiet is a pre-requisite if such results are to be attained.

Suddenly I heard a heavy footfall outside, followed by a punctilious man's voice demanding,

"Ver ist der shithouse?"

The *baba*-seeking German had returned.

At the meal I noticed the Englishwoman from Andorra was missing. The increased cold seemed to make everyone withdraw into themselves again, and nobody mentioned her. We ate in silence, ate some more – and returned to our own private worlds in the chalets.

The swami had apparently changed his mind about me in my absence. As I was leaving the kitchen hut, he remarked,

"If you do decide to trek to Kedar Tal alone, you must remember to cross the ravine high up on the return journey; lower down it is impossible."

I spoke briefly also of my plans to reach Gaumukh and, hopefully, Tapovan across the great Gangotri Glacier. He pursed his lips and looked thoughtfully into the blackness beyond the lamplight.

"You will have no problem reaching Gaumukh, but Tapovan can sometimes be tricky at this time of year. But you can try."

I was impressed by the man's gentle air of inner mastery and would have liked to converse more, but I knew I had done enough talking for the day. A second early night, with the promise of an interesting day to follow was no hardship, and I duly returned to my cabin. During the early hours I was awoken briefly by a gusty wind thrashing through the tops of the nearby fir trees. I listened to it beating down the valley with the speed of a night express and snuggled deeper into my sleeping bag.

Chapter 10

Cloudland

Next morning dawned cold and clear in accord with the swami's prediction. The stars had deserted the sky and the early sun now capped the fretted crests of the icy summits with a glow of resplendent gold; peaks, spurs and frozen notches bristling with ice looked a magnificent sight against a sky of palest lemon. I headed away from the ashram towards the bazaar and Gaumukh beyond, nineteen kilometres higher up the Gangotri valley. A crow called sharply as I passed beneath its perch, and was answered by its mate in another tree nearby. As I crossed the footbridge I glanced into the ravine: the *prana* ball had gone.

Here, if the reader will bear with me, I must make a momentary pause in the ongoing narrative of my journey.

While I am writing this very page at home thousands of miles from the frozen heights of the Himalayas, an unusual and very beautiful thing happens in my room. I pause in my work as a faint rustling sound arises from the already completed pages, which lie in a stack at the side of my desk. I wonder if another mouse has found its way in from the fields, which wear a weary, late winter look

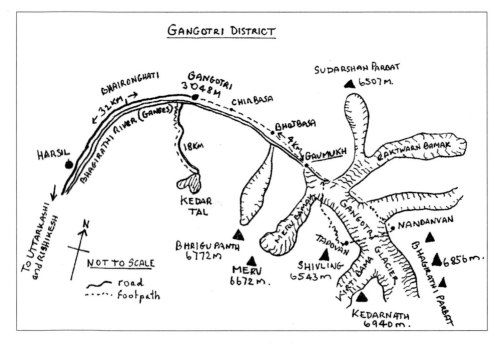

outside my window. Suddenly one of the top pages stirs and I am totally surprised to see a peacock butterfly struggling to free itself. Upon doing so, it flies immediately towards the light and settles on the window pane. I treat the incident as an auspicious moment, even more so when I later discover it hanging upside down from a file marked 'Himalayas'. The event gives me a renewed burst of energy.

The way now opening up before me stirred wild thoughts within me, as if Paradise itself beckoned.

Past the temple was the fantastic background of snowpeaks which overlook the approach to the source of the Ganga. The shapely pyramid of Sudarshan (21,387 feet) was the most striking, but the one I wanted to see most of all, Shivling, the 'Matterhorn of India' lay hidden behind a bend.

In earlier times, the glacier snout reached as far as Gangotri but now the moraine, piled up on either side of the frothing river showed just how far it had receded in recent years (ten miles, in fact). Beyond the temple, a raking incline of steps connected with the main trail hugging the steep, avalanche-prone slopes on the left side of the huge valley.

I could see one or two *babas*' cave-dwellings low down, intriguing hideaways which prompted me to forego the better footpath in order to take a closer look. I assumed I would have little difficulty in scrambling up the banked sides of the moraine half a mile further on. It seemed a fine idea.

Unlike the sackcloth and polythene tenements of the bazaar, these dwellings had obviously been adapted with year-round tenancy in mind. Doors, neatly-rendered walling to draught-proof gaps, and even a tiny glass window in one instance, showed the occupants had made strenuous efforts to seal their homes against the chill blasts and storms of a Himalayan winter. Some were of very singular appearance, often being topped by a saffron-coloured flag or a Siva trident. I peered through the gate of one enclosure to see a delightful, Zen-like garden of simple, uncontrived rock arrangements, dwarf bushes and a paved footpath serpenting to the occupant's front door. All of these interesting retreat-dwellings commanded an inspiring panorama of river, forest and the sublime snow peaks. It was an unmatched situation for those solitary souls who wished to dedicate their lives to the austerities of a higher calling. Some, I had heard, guard against intrusion so assiduously, that they are likely to adopt the guise of a simple ignoramus when confronted by the prying questions of the casual caller.

I might on another day have knocked at the door of one of these contemplatives' cells, but clearly, their locations were chosen because their owners wished to be well-removed from the constant disturbance of the tourist-visitor. To have done so merely in the hope of meeting some miracle-working sage who might know the secrets of astral flight, the Egyptian pyramids, hair restoration and other so-called wonders, which so fascinate the Western mind, seemed to me an idle motive.

I continued along the bush-fringed path which wound its rocky way amongst these troglodytes' dwellings close by the jade-coloured river. Near one garden entrance I stepped aside from what I thought was a discarded blanket on the ground. Only when I happened to glance back some ten yards further on did I notice the steady eye inspecting me from under a raised corner of the tatty rug.

Swami Brahmachidanya had mentioned a man who lived in such a way and I wondered if it might be him. "He is the man I have most respect for in Gangotri," he told me.

There were others with whom he was less impressed.

A short while afterwards I came up to a painted notice: <u>VISIT HOLY MAN</u> it read.

I peered through the gate. According to the swami, this was the home of one who had shut himself away from the world's distractions whilst still clinging to them:

"That man has lived there for forty years in the name of *Siva*. But he is not dedicated to the Lord – he is dedicated to money! From every visitor he asks a donation. In forty years he has collected a mountain of money. All his! But for what purpose!? There is nothing to spend it on. People look after him – they bring him gifts. So he has no need to buy food. He just hoards the money people give him. Now he is old and can hardly walk. When he dies, 'Cash *Baba*', as I call him, will take all his rupees to heaven."

I peered again through the bars of the gate, but could see no sign of him, or, indeed, of a donation box. Perhaps he was indoors counting his fortune?

After this short distraction I decided it was time to regain the main path, whose general line I could discern crossing the shattered slopes above me. It was easier said than done. The trail ended at the last dwelling and there was no obvious line connecting with the higher route. I would have to choose my own way, that much was clear.

Moraine, as every glacier traveller will testify, can be notoriously unstable. The sides of the holy Ganga proved to be no exception. More troublesome than they looked from a distance, the steep inclines of rubble were loose and bore every sign of regular slippage. How to ascend it without mishap? The most feasible route looked to be up a narrowing gully with a large block jammed in its neck about two thirds of the way up. I thought I could by-pass this obstacle without too much trouble.

Accordingly, I scrambled easily enough up the bed of the gully for a hundred feet until I came beneath the roof formed by the chockstone, which weighed several tons at least. Now I began to mount the steeper section, namely the wall on its right-hand side, of cracked, disintegrating granite. Soon, by means of a little deft footwork, I was able to achieve a stance of reasonable security, from which point I intended to climb past the wedged boulder. Three or four moves would give me access to the upper portion of the gully, which immediately sloped back at a more modest angle. It appeared a straightforward enough passage for someone used to airy scrambles and familiar with basic climbing techniques. Two heavy stones were jammed in the lower portion of the drainpipe gap so-formed between the main chockstone and the right hand wall. So long as I avoided them, I should not be threatened from above by falling stones.

In the mountains, however, danger can strike suddenly from a least likely quarter.

At the very moment I allowed my body weight to drop forward and to my right, with my hand outstretched to catch hold above the two stones, the rucksack strap on that side parted without the slightest warning!

The immediate effect was to throw me off balance and with an instantaneous decision to make – if I made the wrong one or hesitated, I had no thick cushion of health insurance to soften my fall when I hit the gully bottom thirty feet below. I had no time to assess the consequences should I drag down either of the two stones upon me – I simple lunged for the largest and wrapped my left arm about it, just as a jerk from the full weight of my rucksack came upon my left shoulder. To my great surprise, the stone held firm, as did its neighbour. Two or three heaves followed before I was safely above the hazard – breathless but safe!

A slithering scramble up a loose slope brought me shortly to the trail, which would have spared me so much trouble had I used it. Upon examination I saw that the strap had not snapped – I had simply failed to notice the buckle coming undone. It had nearly undone me.

It was a dusty start to my ten-mile trek to Gaumukh and showed how easily an innocuous oversight may lead to a significant incident. I took a short breather and a drink, sitting under the spreading branches of an ancient cedar, one of the last to have found a foothold on that part of the mountainside.

The route ahead was not a particularly strenuous one for the fit and able, being maintained in a good state of repair between May and early November, when the Gangotri temple closes for the winter. But in April 1991, two thirds of the way had been covered in deep snow, making the journey a more serious undertaking. Today, however, it was free of snow and few people, if any, appeared to be using it. I was surprised to find it so deserted. To be perfectly candid, I had dreaded the prospect of joining a procession of hundreds, all heading for Gaumukh. I think my fear derived chiefly from a photograph of such a scene, approaching another pilgrim place, Armarnath Cave in Kashmir. But as I have said, the temple would be closing in a week or two and there would be no more buses north of Harsil for six months. Having recovered my composure, I set off again without delay. The path, particularly in the early stages, was a delightful sequence of unexpected views; as one corner was followed rapidly by the next, the magnificent scenery slowly unrolled before me. The morning sun was by now upon me and the cool air was just the right temperature for a steady pace.

Before long, I had left the scattered trees behind and was fast-approaching a more exposed section of the trail, which from a distance resembled a hemline stitched across the lower portion of a two-thousand foot apron of water-streaked slabs. At a number of places where a slip could have meant a hundred-foot fall, the drop was guarded by a flimsy metal handrail. About halfway across I came to a repaired section which I remembered with justified clarity from my earlier visit. On that occasion both the path and the railing had been swept away by rockfall. My Anglo-American companion provided moral support, but crossing the ten-foot gap without the backup of a safety rope was still an interesting experience, to say the least.

A mile ahead, three specks laboured with infinite slowness across the sweeping walls of granite. Otherwise it was as if I had the glacier and towering snow peaks to myself.

Three miles beyond Gangotri I came to Chirbasa. An excellent camping spot amongst Bhoj-patra pine trees, it is used regularly by trekking parties as an overnight halt. Last time I had encountered an Aussie-New Zealand mountaineering expedition there, all Moon-walker plastic boots, bandaged finger ends and unshaven, flaking faces smothered in sun-block. There were no Westerners this time, only two or three tarpaulin structures serving as wayside *chai*-stalls, whose hollow-cheeked owners looked as if they had just surfaced from a cold night. Sunning themselves in the sparkling air were the three 'specks' I had observed earlier, an Indian man and wife and their attractive teenage daughter. They smiled a greeting as I threw off my pack.

Chirbasa, a good place to halt, offered stunning sunlit views of the surrounding mountain scenery. The three elegant Bhagirathi Sisters were now in view, the highest of the triple-peaks being Bhagirathi Parbat (6856m or 22,497 feet). Directly across the river, my gaze was equally taken by a soaring orange wall rising to meet a glistening, and similarly impressive snow dome – one of the Bhrigupanth group of peaks.

I had not been sunning myself for long when I was approached by a wiry-looking man wearing a patched jacket and two days' growth of beard.

"Tapovan?" he asked, looking as if he expected a quick rebuff.

"Yes," I responded, slow to realise his ulterior motive.

"Me guide; I show you Tapovan – glacier very dangerous."

"Me go alone, one person," I countered, "No guide!"

The man seemed not to understand and remained politely persistent in broken English.

I soon tired of the exchange and moved into the gloomy interior of the shelter with my empty glass.

The owner, a middle-aged man, eyed me confidentially as he poured me a second. "He is very poor man with family; he is only trying to earn a little money. If you want guide, you should take him – he is reliable man."

I nodded sympathetically but made no reply. What could I do? Tapovan was at 15,000 feet or more, but it wasn't a climbing expedition to get there as far as I was concerned. And so long as it was free from mist or snow I felt I could handle the Gangotri Glacier on my own. I drank in the views and my tea, paid what was owed and continued walking.

It is a curious but common enough trait amongst mountain travellers to let loose a shower of exaggerated praise when visiting distant ranges, hitherto unknown to them. The gush of adjectives stems, I suspect, from the writer suspending his critical faculties in order to describe what he underlined to see rather that what is there. After all, having travelled so far, and at such cost, one can hardly wait to be impressed! Even when a mountain be no more in appearance than a lop-sided, striated pile – as Everest, it has to be admitted, looks from certain angles – if it already has a reputation for inaccessibility, extreme altitude, and so on, then its lack of beauty, grace and shapeliness of form are mysteriously ignored. Perhaps if the traveller is also a mountaineer he creates his own 'aesthetic line of beauty', a well-proportioned thing in itself, by climbing it?

Having prepared the foreground in this way, so to speak, I have to confess to my own unrestrained response as Shivling, a diamond set in a coronet of lesser gemstones, rose into view in all its sunlit glory. The upper Gangotri Glacier was still too far off for me to pass comment but my first view of this strikingly-formed peak stopped me for words. Though its altitude of 21,446 feet makes it over six thousand feet higher than the Swiss Matterhorn, comparisons made by early European explorers to the region were not, I think, entirely the product of wishful thinking. The Himalayan peak sweeps proudly, awesomely up from its six-mile wide base to a forbiddingly beautiful snow summit; its north face bigger, smoother and more vertical than its European namesake. Even so, when the Matterhorn is observed from its relative altitude, that is, from Zermatt, the six or seven thousand feet of which the actual mountain is comprised, also appears prodigiously steep.

This is the <u>impression</u> one gets. Before Edward Whymper climbed it in the 19th century, rumour had it that a ruined city stood on the summit.

Shivling is more difficult to approach than the Matterhorn and has had relatively few ascents. It exudes an air of mystery and it was not difficult to wonder if some of the higher cliffs concealed a shattered temple or two. The profound silence which clung to it and its shimmering surroundings, causes me to echo the sentiments of Paul Osborne:

"A feeling deeper than expectation, something closer to a spiritual need..."

The further I penetrated the region the stronger the feeling became. An hour or so later, I slowed at a prominent notice in Hindi, a sign which, I had been told, warned of dangers immediately ahead – if the mountain was going to move without warning, this was the place! I took a short halt for a drink, before advancing cautiously.

The high-risk zone was about half a kilometre broad and comprised high-angle slopes cut by steep stone-scoured gullies. Here, from time to time would shoot cannonades of boulders across the exposed footpath, sweeping all before them into the Ganga one hundred feet below. Staring down from high in the gullies stood a division of tottery boulder-clay pinnacles. From forty to ninety feet in height, these bottle-shaped towers looked severely eroded and ready to drop at any time. Numerous accidents had happened there over the years, mainly to people unaware of the hazards of mountain travel. A decidedly menacing atmosphere hung over the area, filling me with the desire to be done with it as quickly as possible. A sadhu had died there recently, crushed by a hurtling boulder.

Accordingly, I hurried with all speed round the windings of the trail, only stopping when I reached the shelter of a granite overhang.

Once beyond that moribund spot, the path took a raking slant across the slopes before finally unravelling itself upon a desolate stony plain ten miles from Gangotri. Another makeshift *chai*-stall marked the intersection of two footpaths, the left-hand branch leading on to Gaumukh and the other descending by a series of rocky zig-zags to Bhojbasa ashram where I hoped to spend the night.

Its appearance this time puzzled me. It was not how I remembered it, a fact which was only explained to me later.

After a short sit down, I set off to enquire about a bed for the night. I was weary of walking by the time I reached the low whitewashed buildings enclosing a sun-parched courtyard and sank gratefully onto a stone bench, tilting my face to the sun. The altitude of 12,500 feet and the dry air may have drained some of my energy, but I had also eaten little since leaving Gangotri.

A few yards to my right, Lal *Baba* (or Red Father), the ashram's founder, sat in animated conversation with several attentive devotees. He was a corpulent yet dynamic figure in his late fifties, and a follower of Vishnu the Preserver, who reincarnated as Krishna and as Rama, the hero of the Ramayana. Apart from a strip of cloth covering his genitals, he was naked. Upon his forehead, three vertical bands painted in red and white indicated his spiritual allegiance. I thought I had better pay my own respects to the master and strolled over in my stockinged feet to introduce myself, an additional incentive being the sounds of activity from the nearby kitchen.

Lal *Baba's* face lit up as he greeted me with a few words of welcome in English. He called out to the kitchen and in a few moments a large tumbler of hot sweet tea was handed out to me.

Shortly, a monk about twenty years his junior briefly joined me as I sat sipping the tea on the bench. He wore a spotless white robe, stood six foot tall and possessed an unequalled gentleness of manner.

"Sit back and rest for the afternoon," he advised softly, "a place for you to sleep will be arranged."

I thanked him for his suggestion, mentioning my previous visit as I did so, when the buildings were surely different to now.

"Ha!" he exclaimed lightly, as if the entire matter had been taken care of in five minutes, "It burned down…we rebuilt it."

Following his kindly advice, I remained on the bench for the next three hours, dozing or occasionally unfurling an arm to reach for my water bottle.

Other than a Hindi phrasebook and a slim volume of *The Dharmapada*, I was carrying no other reading matter. There were good reasons for this. As far as possible, I wished to think my own thoughts rather than other people's. Moreover, too much reading of novels and suchlike was prone to disconnect me from my surroundings; though such abstinence was never an absolute rule it did keep me more centred in the present. I made occasional entries in my journal, but it wasn't a daily event as I believe is often the case with more professional travel writers than myself. Only later would I write things down, recall what was said, and so on. In this way – or so it seemed to me – I would remember only what I really needed to know. My self as a total organism would somehow carry the imprint of whatever experiences befell me, verbal or otherwise. My 'understanding' would thus be less dependent on verbatim notes scribbled down at the time. In any case, there is something absurd to my mind, in the lecture hall habit of feverishly copying down the master's words as they fall from his lips.

By three o'clock in the afternoon I was still the only foreigner in sight. How different is the scene along the popular trekking routes in Nepal, where the impact of visitors on the forests has been devastating. The most visible sign of trekkers passing in those regions is the quantity of rubbish left behind. Puzzled sherpas refer to the sheets of toilet paper abandoned on the trails by Westerners as 'the white man's prayer flags'. But as Swami Brahmachidanya had so forcefully told me, there was also no room for complacency in the Indian Himalayas.

The immediate prospects for the following morning appeared much better than in 1991. On that occasion the prevailing weather prevented me from going beyond Lal *Baba's* ashram. Arriving cold, tired and with feet and boots soaked by wet snow, we were shown every hospitality during our short stay. The monk who attended us spoke perfect English and informed us that almost all he now owned was on his person – a blanket and a pair of cumbersome-looking boots. The rest he lost when he had to leap from the window of a burning bus, which had been attacked near Haridwar. Seven passengers died and he counted himself fortunate to have escaped unscathed. But, strangely, he kept us waiting outside in bare feet. With the temperature close to zero, I was puzzled as to the reason. After half an hour, however, he showed us indoors and sat us before a tin-can brazier fuelled with honey-coloured wood shavings.

"If you had come immediately before the flames," he explained, "it would have been very painful for you with cold feet."

Fortunately, there was no competition for the fire's modest heat since we were the only ones to have attempted the trail that day. While we were thawing out, I noticed my companion had a polythene bag stuffed with four or five spare pairs of thick socks. My only pair were soaked and would probably remain so until I put them on again in the morning. Swallowing my pride, I asked if I could borrow a pair of his. We weren't close friends, more companions of convenience, but we had overcome a tricky situation or two together during the trek – I couldn't see any problem in using a pair of his as bedsocks. I was wrong.

"No, I'm sorry," came the unexpected reply, "I need them."

I can laugh about it now, but his response surprised me at the time.

Later on, he loaned me a hundred rupees on the return journey. I was grateful for that.

Our quietly unassuming host of that visit was another of those independent souls whom it has been my good fortune to meet whilst wandering the Himalaya. In his mid-forties with blue-black shoulder-length hair he was, he said, a writer on philosophical matters and lived for six months of the year at 6000 metres because only in such high, secluded surroundings, did he receive the inspirations of which his writings comprised. Upon spiritual questions he spoke with impressive authority, yet with a becoming modesty which made one more inclined to believe what he had to say was true. He was also a collector of medicinal mountain herbs.

When we parted company, he presented me with a simple gift – a small package containing a quantity of these rare herbs which he had personally collected above 4000 metres.

"Whenever you are short of energy on your travels, you should inhale," were his parting words. It wasn't *ganja*.

This time the snowline was much higher – but for how long?

I squinted again at Shivling (or 'Ganesha's Conche' to give it its other name). Looming as it does over the untainted Sanctuary and the Ganga's source, its steely ice cliffs and fabled granite walls which protect the summit, have long attracted many top class mountaineers in search of a supreme test of their skills. But it was not until 1979 that the Indian authorities re-opened this dramatic massif to Western exploration. Much energy has been expended in achieving some notable ascents of the various routes up the mountain.

Perhaps the mountaineer and the 'sedentary' yogi this much in common:

Both have to cope with solitude; each chooses to give up much in the way of personal possessions, material security and position in worldly life. An inner calling has to be answered and for a few weeks or months at least, the former leaves irksome routines of everyday life behind in order to fulfil an ambition. The monk or nun who enters upon the monastic life is prepared also to part from friends, family and treasured belongings in order to travel further upon the untrodden way of personal development. They too embark on what may well become a lifelong expedition, hoping that one day they will reap the fruits of their labours. The tricky part for the mountaineer is to carry back to the world and use the lessons he has learnt in the mountains. If he can't, then for him there is but one answer. The monk is likely to deny that he ever left the world: he carries it into the monastery with him.

In my little book of wisdom I read the following journey-stopper:

"The further you go the less you know. Thus the sage knows without travelling; He sees without looking;" (Tao Te Ching)

When the sun dropped suddenly behind a ridge to the east of Shivling peak, the cold pimpled my arms as if a giant refrigerator door had suddenly swung open in front of me. A chill rapidly filled the air and I hurriedly pulled on my crumpled shirt and pullover. Lal *Baba*, rolling upright and covering his ample body with a blanket, glanced across:

"You should get a bed now – tonight very cold!"

The monk who had assisted me earlier waited while I repacked my scattered belongings, before leading the way across the courtyard to a door in the single-storey accommodation block. Within was a passage, lined with wall-to-wall polythene carpeting and a number of bare-looking rooms on either side, their floors covered by mattresses of dubious pedigree. Indicating I should enter the first room on the right, he then left me to my own devices.

Though the room was empty now, I knew from past experience that evening would frequently bring a rush for space and accordingly, I claimed a place against the wall furthest from the door by unrolling my sleeping bag against it. A number of quilts and army-type blankets were scattered about and I used one of the latter folded up as a pillow. After pulling on the remainder of my warm clothing I returned to the now chilly quadrangle, where the monk in white appeared to be star-gazing in the fading light. He quietly mentioned that now was a good time to square-up for the night's lodgings.

"*Kitna Hai?*"

The reply was no less assured:

"Ninety one rupees."

I gave his face a double-take – why ninety <u>one</u>, of all numbers? Perhaps it was an opportunity to show generosity by foregoing my change? I looked for a twinkle in the monk's eye as I handed over a one hundred rupee note, but there was none.

Surprisingly, the change was immediately forthcoming, being handed over without the slightest flicker in the bearded monk's calm expression.

The evening meal was still two hours away, so I decided to take an exploratory walk around the boulder-strewn plateau to warm myself up.

Nothing moved on the grey acres of stones as I set out. Above the valley's armour-clad rim, the surrounding peaks were still flushed by subtle tints of pink and gold, whilst a faint breeze roamed like a lost spirit amongst the silent stones, stirring memories of other times, other places. Each hillock, each stone seemed only to exist for one purpose, to affirm: I AM. As I crouched amongst them, a solitary star glinted like a pale gemstone, low down in the western sky. I watched the play of cold fire slowly fade before it occurred to me to unlock my cramped limbs.

In *Mount Analogue*, René Daumal explains the symbolism of mountains for mankind:

"In the mythic tradition the mountain is the bond between earth and sky. Its base spreads out into the world of mortals whilst the summit reaches into the sphere of eternity. It is the way by which a man can raise himself to the divine and by which the divine can reveal itself to man".

This is nothing new. For the Greeks, Mt. Olympus was revered as the lofty dwelling-place of their deities, whilst in the Far East, Fujiyama is held in similar esteem by the Japanese, the Mt. Kailas of Japan. In California there is Mt. Shastra.

Hindu mythology has woven numerous stories around the sacred peaks of the Himalayas. On the platform of every Hindu temple stands a *shikhara*, or altar, which by its architectural form succinctly denotes the inner path or mountain that each of us, in our own ways have to climb on our life's trek. The mountain's component elements – height and shape particularly, are readily transposed to the lofty 'other' world of spiritual ascent where we realise our true nature and connectedness with all things. The 'body' of the mountain may be interpreted as spirit made manifest; its vertical axis corresponds to the planetary axis and, on the human scale, to the spinal column. This notion of mass united with the idea of verticality expresses a profound symbology which is common to almost all traditions. As with the image of the Cosmic Tree, the location of this sacred mountain is at the 'Centre' of the world. Perhaps arising from deep-rooted human longings, the terrestrial location of this mystical peak seems to be a matter for individual preference. There is a Mount Meru of gold situated at the North Pole which links neatly with the idea of the centre and the Pole star, the 'aperture' through which all corporeal elements must pass before the soul can enter the Subtle World of Liberation. The Gross physical world of Time, Space and Elements must, according to ancient Yogic tradition, be transcended if immortality is to be realised. Marked on my map was an actual Mount Meru (6672 metres) which may be reached via the Meru Bamak Glacier, a narrow tributary of the Gangotri Glacier. The anatomical correspondence to this worldly axis is the *Bindu*, the point at the top of the back of the skull. It is not a physical place so much as a tiny area where psychic energy and physical function meet, a *chakra*, often symbolised by a crescent moon or stars.

Milton appears to have contemplated *Meru*, "the spiritual and phenomenal centre of our world...the axis which connects earth with the universe, the antennae for the influx and outflow of spiritual energies":

> Against the eastern gate of Paradise
> Leveled his ev'ning rays. It was a rock
> Of alabaster, piled up to the clouds,
> Conspicuous far, winding with one ascent
> Accessible from earth, one entrance high;
> The rest was craggy cliff, that overhung
> Still as it rose, impossible to climb.

(Paradise Lost, Book Four)

In Celtic and Irish stories the mountain is often described as hollow, a place of both the Living and the Dead. Legends from Asia, Europe and elsewhere, frequently tell of a hero or heroine asleep inside the mountain, one day to emerge in order that the opposing poles of life be reconciled.

Viewed from above, the mountain, gradually widening to its base, corresponds to the 'upside-down tree' of Eastern mythology whose roots reach towards heaven while its leafage expands downwards, thus signifying the universe in its constant rhythm of expansion and contraction from the One into the Many. As Eliade suggests, 'the peak of the cosmic mountain is not only the highest point on earth, it is also the earth's navel, where creation had its beginning'. The Mount Meru of Indian legend is said to be a seven-sided pyramid (corresponding to the seven planetary spheres, the seven Virtues and the seven Directions). Each face is a different colour of the spectrum. When seen as a whole (like a multi-coloured spinning top) its colour is brilliant white, by which account it may be linked to the image of Totality or Oneness. As a landscape it can be regarded as a mandala in which all paths lead towards the Centre.

For both the Tibetan Buddhists and Hindus Mount Kailas is regarded as both the highest and noblest of pilgrim destinations – I referred to the significance of the place in an earlier chapter. Known in Tibet as *Gang Rinpoche* 'Precious Snow Peak', its summit by all accounts is a most crowded perch for all manner of deities. Malarepa, the great yogi who attained the top after a contest in magic with a shaman priest of the ancient Tibetan 'Bon' religion, lives there. Lord Buddha along with five hundred *Bodhisattvas* is also said to be in residence on this most sacred of summits. From the west, according to some who have been there, the mountain resembles an enormous, threatening cobra. Both Hindus and Tibetans alike perform the traditional *Parikramas* (circumambulation), an arduous undertaking of two days including an icebound pass at an altitude of 18,600 feet. One fifty kilometre round is said to wash away the sins of a lifetime. A glance at the map will reveal that Kailas, at over 22,000 feet, is the apex of the 'Roof of the World', as the high Tibetan tableland is called; moreover, the four great rivers, Brahmaputra, the Indus, the Sutlej and the Karnali all have their source in the Kailas-Lake Manasarover region. Lama Anagarika Govinda writes:

"Its shape is so regular as if it were the dome of a gigantic temple, rising above a number of equally architectural forms and bastions and temple-shaped mountains which form its base... And as every Indian temple has its sacred water tank, so at the southern foot of Kailas there are two sacred lakes, Manasarover and Rakastal, of which the former is shaped like the sun and represents the forces of light, while the other is curved like the crescent moon and represents the hidden forces of the night".

The tantalising fact was that, as the crow flies, I was a mere hundred miles south of the place which Govinda describes so vividly. Be that as it may, the drop in temperature and the sun's slow descent behind the western ridges soon steered my roaming thoughts round to matters nearer home and I continued my stroll, soon finding myself nearing the draughty, but hospitable canvas *chai* shelter. Within, four weary-looking Westerners, including the saturnine Italian, sat sipping hot drinks. The others were two young Englishmen only four days out from Delhi airport and a plump and unfit Israeli woman from Rishikesh. We exchanged travellers' information in desultory fashion. They had spent six hours on the trail and would be sleeping the night at the grandly-named Tourist Rest House, a bleak concrete block which looked more in keeping with a W.D. depot than the foot of such a magnificent mountain as Shivling. In the morning they intended to carry on, but only as far as Gaumukh.

I focused on warming my hands from the heat of my tea tumbler. The Italian seemed less diffident than before, telling me he was from Turin and how difficult it was to arrange time off from work to go trekking. I was glad for his sake that he had found company, though my motive, to be honest, wasn't entirely selfless. When my glass was empty I rose to go.

"Maybe see you tomorrow?" the Israeli woman said.

"Perhaps," I replied, raising a hand in goodbye.

I could see my breath misting in the cold air as I made my way slowly down the bouldery zig-zags.

H.S. Shivaprakash wrote in *Stones*:

> Strong hands rain
> blow after blow
> on the hearts
> of gigantic stones.
>
> ....
>
> But only after taking
> such heavy beatings
> and incessant thrashing
> does the stone become
> tender like the heart
> and the heart
> hard like the stone.

In my absence, some newcomers, four students from West Bengal, had turned up at the ashram. All were friendly young men, full of high spirits and only too eager to demonstrate their command of the English language upon me. Hovering on the edge of the group was the local guide who had offered me his services earlier in the day. When I returned from answering a call of nature, he was in earnest conversation with three of the students. Suddenly the most talkative, Ajay, swung about and addressed me eagerly:

"This man is guide. He says he will take us to Gaumukh and Tapovan for four hundred rupees – you can come with us! It is not much to pay."

"It's all right," I replied affably. "The money isn't the problem, but I prefer to go without a guide."

Ajay, clearly concerned about my safety, was not to be deterred so easily, and proceeded to warn me of the perils that lay in wait for the witless foreigner wandering alone in the mountains.

"On the glacier it is very slippery...there are crevasses. If you don't know the way it is easy to get lost, fall down, This man – he will show us - " And so on and so forth. He seemed genuinely surprised that I didn't leap at the opportunity to enlist in his jolly band. I had heard all this before.

"Look," I said patiently, "I am not going up there to die! At this time of year the glacier will be dry – all the crevasses will be visible. There will be no problem with the weather...If there _is_ cloud, I know when to turn back."

Ajay still looked doubtful, but his friend Indranil, who had said little, broke in quietly:

"I think this man can go to Tapovan alone – you will have no problem without a guide."

"Yes, yes!" the third student gushed, "_you_ can be our guide!"

Everybody laughed, myself included. Everybody that is, except the real guide.

But I didn't let the subject drop without making it clear that although I enjoyed their company, it was best that the students didn't rely upon me for their safety. Ajay pondered the remark for a moment before addressing his reply to the others:

"Yes, that is right; you are right."

After a pause, the fourth student, who looked paler than the others, complained of feeling unwell and said he was going to lie down in our room. I remained outside for a while longer, bending my ear to Ajay's not uninteresting views and opinions on a blitz of national and global topics. He appeared very well read and launched into a rapid-fire survey of headline issues:

"Everywhere I look in the world people are depressed! Why is that? Why is there so much widespread depression!?"

"Er," I began sagely enough – but before I could continue, he was on to the next burning theme:

"In your country, there is Northern Ireland – I read about it in Indian newspapers."

"Yes," I ventured, "a situation with no easy answer. I - "

Again he raced ahead of me. "Yes, you are right, the Irish situation is not just a British-Irish problem: conflict is a world-wide question!"

At first his enthusiasm carried me along, but after a while my attention began to drift. Hoping to steer the conversation away from hard-boiled politics, I launched the opinion that, "unlike most of his profession, Nehru had a very marked poetical, if not religious side to his nature."

Ajay shook his head like one who knew the truth of these matters: "No, no, that is not correct! Nehru was a politician – he only used religion for his own political purposes."

"That may be partly true," I replied. I had already noted that the short word 'grey' did not figure too prominently in my new-found friend's vocabulary and I let the subject drop. But I had underestimated Ajay too hastily.

Pouncing on my use of the word 'poetry' he asked me whether I had read, written or published any. Which Indian poets did I know? (I mentioned Tagore) Were there any poetry magazines in England which would interest him?

"Many famous writers come from West Bengal," he enthused, "but it is very difficult for them to be known outside India – there are problems of translation, you see." He looked at me as if I might be a publishing house translator, poet and roving talent spotter all rolled into one.

Good God, two days ago I had been hailed as the conquistador of Shivling, while today I am considered to have my finger on the pulse-beat of literary England! I had better curb my ingenuous disclosures from now on, before the pit I have dug for myself becomes too deep to climb out of.

Fortunately, the fast-falling temperature provided me with an excuse to break-off the conversation – although only temporarily I suspected, an intuition which proved to be well-founded. I took to my sleeping bag for a rest. Though I had been taking copious amounts of liquid since reaching Bhojbasa, I had more of an appetite for a good night's sleep than a full ashram meal; tomorrow promised to be another long day and all my body needed was a few hours respite. My head had scarcely touched the pillow however, when I heard a soft tread on the crackling carpet outside my room. An oil lantern held aloft, swung in from the blackness of the passage and the monk in white called, "It is time to eat."

I told him I wasn't hungry. But when he gently insisted – "it is necessary because of the altitude" – I thought he might have a point, and began to struggle back into my outer layers of clothing.

Seemingly because of my previous visit to Bhojbasa, I was invited straightaway into the snug lamplight of the low-ceilinged kitchen. A smile from one of the two cooks flashed a welcome across the bubbling pots, and I sank gratefully to the floor, glad not to be waiting outside in the darkness and cold. The pale lamplight drew the four of us into a circle, a group in which I felt immediately at home, although I was unable to penetrate much of the occasional exchanges. The man opposite was a Nepalese who spoke a little English. He worked also as a trekking guide, he told me, and we talked briefly on that subject. In truth I was tired and wasn't sorry when the conversation tailed off into a reflective silence which felt more appropriate to the way I was feeling. Eventually the monk, who had been ducking back and forth through the low doorway, handed me a tin plate swimming with cooked *dal* and two chapatis. There was no wine list, but a large tumbler of scalding chai helped appease my constant thirst. Both chapatis were stone cold, I was surprised to discover . Shortly, when a third was offered me direct from the pan, I decided there was no point in suffering for politeness' sake and stuffed the two cold ones surreptitiously into my pocket for future disposal.

It wasn't long before the combined influence of kerosene fumes, food and tobacco smoke began to take their toll, and a comatose state began to settle upon me. I scrambled outside for a final look at the night sky. The pure, chilled air cleared my head, but I paused for no more than a moment or two to observe the twinkling dome, before diving once more into my sleeping bag. It was a few moments after eight.

With five of us now crammed shoulder to shoulder as if in a sardine tin, I could not be accused of hedonism in my choice of overnight lodgings: I leave it to the reader's imagination to picture the shortcomings of such an arrangement. As luck would have it, Ajay was on my immediate left, with his sick friend in the middle, lying with corpse-like stillness. Of course Ajay saw this as a golden opportunity to continue our earlier conversation and this he did with a warm-hearted enthusiasm difficult to resist. At first I answered him intelligibly enough, but when he was still speaking with the same fervour forty minutes later, I yawned theatrically and told him I was now going to sleep. He must have been keeping the others awake too, for a peeved voice cried out in the darkness:

"You are talking too much, Ajay – let him sleep!"

Ajay chuckled softly and whispered, "Yes, they have heard my speeches before! But you must have my address for when you are next in Calcutta."

I rolled thankfully on to my right side with my nose no more than six inches from the cold stone wall and, as stillness descended upon the room, dropped immediately into a deep sleep.

Some hours later – I could not be sure how many – I woke up to the sound of urgent whispering. The room was as black as the gallery of a mine, but someone was astir. A few seconds later I heard a creak at the door, followed by shuffling footsteps receding towards the exit. Grim retching sounds accompanied by groans, reminiscent of my own student days, soon reached my ears through the black PVC window. A silence followed before the footsteps returned, to be followed by much shivering and the disturbance of somebody settling down again. It was Ajay attending to his sick companion. This procedure was repeated twice more during the next two hours, a disquieting mixture of stealthy movement and tortured sounds which did nothing to help my own condition: I too had awoken with a fragile feeling in my stomach.

Porcelain bathrooms are not yet on the agenda for such basic dwellings, but I was reluctant to heave myself outside into the night. Instead, I experimented with different lying positions in an attempt to combat the feelings of nausea: flat on my back was definitely the least helpful. I also tried to keep negative thinking at bay by visualising the affected part of my body as "well". The repetition of affirmative, mantra-like phrases repeated silently over and over again seemed to help, and presently I fell into a period of lighter sleep. I suppose two or three hours must have passed before I again returned to the waking state. Overhead the sheet metal roof which had been hidden in darkness for so long, was now pin-pricked by a dozen points of light whilst the covered window, announcing dawn, had begun to change colour. My sickness had all but vanished, but its legacy was a taut band across my temples. For a while I lay motionless, listening to my own breathing and wondering if my headache was simply the result of sleeping in a confined space or the first signs of altitude sickness. The grey light did little to accentuate the non-existent charms of the bare room. As soon as I cleared my throat Ajay rolled on to his back staring listlessly at the ceiling.

Hearing that I was awake too, his face swung towards me. "I am unwell this morning," he moaned sadly, as if he had somehow failed in his duty towards me as a visitor to his country. "The food is not good here," he continued in the same vein, "the water, also." The others were in a similar state to himself, he said, and none of them would be well enough to start the final trek to Gaumukh, not to mention Tapovan.

I sighed —whether to go or stay? But the brightness of the outside world called to me and I began to struggle from my sleeping bag. To linger would in all probability mean succumbing to the general malaise and I might never get another chance. I squeezed into my cold boots and staggered outside, thinking a change from breathing the confined air could only help the state of my head.

It was not yet seven o'clock and the crisp air which greeted my exposed extremities told me the temperature was still several degrees below freezing point. Lal Baba was nowhere to be seen, but the monk who had received me stood in rock-like silence, his face to the east, as though waiting to salute the sun. I had the impression that he didn't need as much sleep as most people. Sensing my presence, he turned to greet me in customary fashion, his face quietly alive with serene thoughts. He ducked inside a doorway for a moment, emerging to hand me a

239

scalding chai. While I was supping the early morning lifesaver I told him where I was going.

The monk's response was instant and decisive:

"Now the sun is nearly up – if you are going, you must go now!"

Bhagirathi's lofty summit ridge resembled a line of frosty sails caught in a frozen sea, but dawn was now advancing long beams of gold across the lonely cols and through gaps in the walls of stark ice and rock; most of the peaks had begun to lose their inhospitable wasteland look. Far off, as if in another world, Shivling's proud summit gleamed in the early sunlight like some dazzling fairy castle – with such a sight to motivate me I needed no encouragement.

A quick goodbye to the suffering students and I was on my way scarcely half and hour after rising.

After only fifteen minutes steady walking my headache had totally vanished, a fact which eased any lingering doubts that I might be about to succumb to altitude sickness. The air though cold was perfectly still, and so long as I kept moving steadily I was warm enough in my minimal clothing even at that early hour. No-one was astir at the charmless Tourist Lodge, but I could see two slowly moving figures fifteen minutes ahead of me on the slanting six kilometre footpath to Gaumukh. From the size of their loads and unhurried bearing, I guessed the figures to be porters or guides. Due to a storm and deep snow I had not even got this far on the last occasion.

In summer conditions it is an easy-enough walk, but in the past, when there had been no fixed trail, travellers had to listen for the rush of the river if it was snow-covered and follow the sound. Today, its line is marked by stones painted red and white and is a simple matter to follow, providing the conditions are favourable, which can never be assumed at these altitudes. According to what I had heard at the ashram this final lap of my trek would take around two hours. But when I topped a rise after only an hour's brisk walking I spotted two patched ridge tents drooping amongst boulders only a short distance ahead. This canvas outpost was not Gaumukh as such, I quickly realised, but marked the end of the trail as far as easy walking was concerned. Ganga's Mouth lay a little further on across a jumbled moraine over which I would have to pick my own way.

Two sleep-bleared faces stared blankly out from one of the three-sided shelters when I came to the tiny encampment. A kettle was boiling gently in the background, but I decided to continue on to the ice cave before calling a halt. By way of encouragement, I gestured behind me, "more people coming."

The way ahead now lay over the chaotic heaps of moraine boulders with many a hole for the unwary to stumble into. From a distance they bore a marked resemblance to currant puddings, between which were gravel runs and the ultimate results of a glacier's colossal grinding action, a fine grey powder resembling builders' cement. Would the mountains themselves be one day reduced to this I wondered, as I scrambled over the football-sized rocks. An eagerness stirred within me as the nearby dash of the new-born Ganga grew louder…

Proceeding forward a few yards more along the ice-rimmed edge of the onrushing water I noted an occasional boulder painted with the familiar red stripes, the odd scrap of bleached cloth, even a discreet Siva trident. Otherwise there was nothing to indicate I was approaching this 'sanctuary of Sanctuaries' amongst the world's highest mountain range… A scuffed toe-print or two, a bird-scattered remnant of an offering to the gods – and I was there.

Without warning, a large vaulted cave, its colour varying shades of glistening blue, green and white was suddenly before me: I had reached the Cow's Mouth, the source of the Ganga.

I stumbled forward and stopped, staring into its arched interior. The cold and all tiredness left me in a flash. This _was_ the place.

Words begin to tumble over each other in an attempt to convey the meaning of such moments afterwards. It was exhilarating, humbling and ordinary, all at the same moment. I felt the atmosphere of ancient sanctity, of spirit and light with which this place had been blessed. The freezing turquoise water bubbling from the cave's sculptured interior seemed to carry with it a timeless echo of those devouts, too numerous to count, who had once stood where I was now standing. We had all done the same walk and in so doing we each helped perpetuate the Way for others who might follow in our steps. Arrival was humbling because I remembered the importance of Mother Ganga to countless millions of Hindus, a living symbol, 'the soul of India which has watched the events of Indian history as a silent observer through the ages'.

Of course it was impossible for the source of the Ganga to have the same meaning for me as it had for those born and nurtured in the Hindu culture: I could not even conduct myself like true pilgrims do on such occasions, only too eager to offer their prayers and thanks at their journey's end. In 'Garhwal', (Ed. A.P.Agarwala), a photograph has the touching caption:

'In this prismatic shimmering ice-scape of snow and ice, pilgrims stripped to the waist take a dip in the foaming, silvery river singing its way through the sculptured rock of the deep piercing gorge while sadhus sit in deep meditation'.

The silence of the 'ice-scape' would always be there. It is only we with our chattering minds who lose it.

Had it been later in the day I might have followed the time-honoured ritual and waded out into the current for an obligatory dip, but some hours would have to pass before the sun's warmth would send blocks of ice crashing from the greenish-white walls of the cavern's mouth. My finger tips and nose were already pinched by the freezing air, so I opted for the coward's way out and decided to take a mere leg dip instead. Having come this far, I felt such a gesture done in the right spirit was called for before I moved on.

I should have been prepared for the frigid temperature of glacial melt-water, but I caught my breath in shock as soon as I lowered my bare feet into the swift-flowing Ganga. A few seconds was as much as I could stand before I withdrew to the bank, where I set about my feet with my much-travelled towel.

William Blake, the eighteenth century mystic, wrote:

'Great deeds are done when men and mountains meet,

'They are not done by jostling in the street'.

He was right. I cooked a packet of vegetable soup, heated over two solid fuel tablets which I carried with me for just such an occasion. While I was at it, I also ate a curling cheese sandwich and a couple of ounces of dried fruit, trying my best to think of it as not merely fuel for a wolfish appetite but as an offering to Mother Ganga and all other lordly deities who presided over the region.

The majority of pilgrims go no further than Gaumukh but I looked once more into the seething throat of the ice-cave, then decided: I would head on to Tapovan – if the gods were satisfied with my offering of a cup of Knorr's soup and a three-day old crust then, hopefully, I wouldn't be claimed by one of the glacier's many crevasses. What was it about that speck on the map which drew some holy men to spend an entire winter shut off from the world at 15,000 feet? I wanted to find out.

It was nearly nine o'clock when I began to mount the steep slope of moraine debris to the left of the 100-foot ice wall containing the cave. Footholds were nowhere secure, every step an exercise in uneasy equilibrium; each time my foot slipped a mini-rockslide would threaten, causing me to pause whilst the stones stopped their downward grumbling.

After progressing in this way for 200 feet I arrived in a perspiring condition at the Nandanvan-Tapovan footpath, where I immediately threw back my balaclava and put away my gloves.

The main track, if it could be called such, continued up the left side of the glacier towards the Bhagirathi peaks; my way, however, lay to the right, a winding course across two kilometres of jumbled ice ridges rearing up amongst a sea of crevasses. In its lower reaches at least, my immediate impression was of unmemorable terrain, an undulating river of chaotic rubble and gritty, ill-defined ice waves whose frozen sides were so peppered with stones that I almost forgot I was crossing a slowly-crawling Leviathan some thirty kilometres long. Rock slides from the ranges on either side had spewed a carpet of moraine across the glacier's surface, a covering which made for easier walking, but was not a pretty sight at close quarters. In its defence, I am sure that the first winter snowfall would utterly transform the scene which in late October had none of the overwhelming grandeur of other glaciers on which I have had the good fortune to set foot.

Initially, a faint line of footprints following a succession of low stone cairns pointed the way. When the former petered out I gave up looking for them and tried to judge for myself which was the simplest way. Now and again the ice shone through in the beds of this labyrinth of V-cut troughs – I hesitate to call them crevasses since none were chasms of soul-searing depth ready to swallow an incautious step. Route-finding, especially in mist would be the greater difficulty.

To judge from guidebook sketchmaps one would imagine that such a crossing would simply involve picking a landmark on the far side and setting out for it in a direct line at a lively pace. Glacier crossings are rarely so simple. By the time I drew near the central portion, the ice ridges had not only grown considerably higher, but very inconsiderately ran in several different directions at once. Scrambling amongst the troughs of these frozen waves I could now no longer see the cliff on Shivling's lower flank which I had picked out earlier as a marker. I hesitated before the assortment of possibilities. The sun had now risen above the mountain barrier and I was wearing my snow goggles as a precaution against snow blindness and the splitting headaches which invariably accompany that plight. All was still in that weather-eroded place and I could see no sign of any human being.

Suddenly a piercing whistle rang out!

I looked up and across several meandering ridges to spot three men, probably the porters from lower down; they were resting with their backs set against heavy, compact loads. The eldest swung his arm in an arc to indicate I should be more to my right. I was a few degrees off the proper course it seemed, and I gave an acknowledging wave in return. When I caught up with them ten minutes later, I discovered they were ferrying kerosene to a Malaysian expedition base camp beyond Tapovan, which had nearly run out of fuel. "No running water, so they have to melt ice," one of them explained with a grin.

On I went.

I reached the edge of the glacier with no further incident and immediately began to pick my way up a very steep and tortuous slope of crumbling, straw-coloured rocks. It was, in fact, the base of Shivling's west face. Far up, a forbidding distance above me, a minute orange flag hung from a dark crested ridge as if in mockery of the puny figures toiling below. This was my immediate destination.

As I paused to estimate how long the climb would take me, the sun flashed on something bright, perhaps a wristwatch or a pair of goggles, a short way below the flag. A second later I spotted two tiny figures descending, one moving steadily and with confidence, but the other edging down with painstaking slowness. I followed their progress for a few seconds until they disappeared one after the other behind a protruding cliff and were lost from view.

As I slowly ascended the mountainside, a bird's eye view of the Gangotri Glacier unrolled beneath me. It was a case of 'distance lending enchantment' to that section I had just negotiated, but the view of the upper portion, sweeping away into the distance to vanish in a curve of unblemished grace within a cirque of 20,000-foot peaks, was compensation of the highest degree. On the furthest horizon ice ridges shone nobly in the glittering light, a silvery tightrope stretched between earth and heaven across a sky of azure blue.

Now I was beginning to feel the first effects of the altitude and every eighty paces or so I stopped for a few seconds to catch my breath. Perhaps it was the lateness of the season, but I was expecting to see more people, particularly foreigners. So far as I could judge in a brief encounter, visitors were not yet causing the kind of problems which the neighbouring country of Nepal has to deal with. In that country a saying has arisen amongst those who care about the natural environment: 'Now we have three religions – Hinduism, Buddhism and Tourism'. The Gangotri yogi had sounded a warning, but how long before similar such sentiments are expressed about the Indian Himalaya, is anybody's guess. Bhutan has taken heed of what has happened across its own mountainous border.

After I had been labouring upwards for the best part of half an hour, I heard voices overhead.

I stopped just in case as a clatter of dislodged stones shot round a corner. A few seconds later, a thirsty-looking pair, an elderly man, his face lined by fatigue and a bearded, much younger companion, appeared fifty feet above. It was the Everest climber whom I had met on the bus. Delighted that our paths had once more crossed, we halted to talk for a few minutes.

He was determined to introduce me to his companion with the same degree of misrepresentation as before, adding for good measure, a salutary nod in the direction of Shivling's remote summit, still some 6,000 feet above us. A weary smile crossed the older man's face: Clearly he was in no fit state to be impressed by my distinguished credentials. After Gangotri the pair were intending to trek some more around Badrinath, another justly famous region a day or two's bus ride away. We all had our destinations to make, and in a short while I left them to continue their slow descent to the glacier. I forged on, dominated by the notion of reaching the ridge before taking another proper rest. Shortly after midday I at last reached the flag where I threw off my rucksack and sank thankfully onto a convenient stone.

244

The Tapovan plateau now undulated before me, no nondescript boulder field, littering the austere western flank of Shivling, but a golden meadow with an ice-fringed rivulet meandering quietly through it.

The name 'Tapovan' means 'meditation ground' and earlier in the season the peaceful hollow would have been a luminous green, sharply set off by the wild grey rocks around it. The place is popular both with herds of *Bharal*, blue mountain goats, and the occasional expedition base camp. I can't say I was too upset at the absence of a nylon village of carbon-hooped tents and piles of mountaineering equipment; there were no goats either, which <u>was</u> a pity.

I didn't know what to expect in the way of resident hermits, but proceeding forward I soon came in sight of a low, stone-built dwelling standing inconspicuously on a rock shelf to my left. This was where Bengali *Baba* lived, the Everest climber had forewarned me. I was about to pass by when a voice rang forth from the dark interior of the open doorway:

"Hello – welcome! Come, come!"

The hut's premier situation had an obvious strategic advantage for the *baba* in that all who visited Tapovan would walk by his door before anyone else's. A face and a beckoning arm appeared momentarily at the only window to further encourage me.

In some parts of the world a low entrance is a common enough sight wherever people live, a feature which I had long assumed to serve the practical purpose of repelling the weather. The indigenous population might well be unusually short, but there is one further explanation.

Hindu culture, for example, teaches not just respect for the parents who have brought one into the world, but also for the house that has given protection from the rain and the sun and from the harmful deities outside. The lintel of the outer door is thus traditionally kept low so that he who enters the house is forced to bow down as he does so.

I am sure I must have presented an appropriate spectacle of grovelling servility as I dropped to all fours in order to pass through the hatch-like opening. In the gloom I almost banged heads with the porters who were just emerging. My eyes took a few seconds to adjust to the dim interior.

"Please...Sit!" a strong voice commanded as I floundered onto a piece of sackcloth with my back to the wall.

Bengali *Baba* – "a man of principle" – as the Indian mountaineer had described him, sat facing me on a dais of folded rugs and cushions. He was wearing a flimsy cotton garment more suited to the beaches of Goa than the ice-bound wastes of 5,000 metres. Probably in his early forties, his black hair, thinning above his brow, fell about his shoulders to frame a broad, intelligent face. Below was a beard

245

containing a thread or two of grey. Throughout the entirety of my stay his legs remained neatly crossed in the lotus position. The illustrious *baba* fixed me with a warm but unblinking gaze.

By now I had grown accustomed to such treatment from total strangers in India and returned his look with equal interest. I had heard more than once the claim that yogis of very advanced spiritual development frequently have eyes of a remarkable colour. With a jolt I realised that the unwavering eyes returning my stare bore this arresting feature of being red. Not bloodshot red from a late-night hangover, but one of penetrating depth, of a deep ruby or garnet hue. It was easy to be impressed. Even in the rarefied ozone layer of Truth seekers, there are sufficient impostors about to make one cautious, but I felt that here was a man who might experience those spiritual realms about which others only dream. Would he convince me?

Bengali Baba obviously liked the company of visitors and questioned me in sparse English as to my origins and purpose in visiting India. Our exchange quickly steered itself to a subject of mutual interest – the power of high places and the value of silence. During one of the long pauses which punctuated our talk, he suddenly called out in Hindi towards a nearby room. A stove spluttered into life and shortly, his helper, a woman a few years younger than himself, emerged to hand us each a metal tumbler of tea.

"I am enjoying our conversation," Bengali Baba announced candidly as we sipped the scorching liquid in silence.

I was too, but in truth I was beginning to sense that our conversation would soon begin to founder on an insufficient understanding of each other's language. How to extricate myself before conversational platitudes began to take hold? As though thinking the same thought the *baba* observed, in a mildly chiding voice, "If you come to India, you must learn Hindi!"

A sufficient pause for this advice to sink in followed, before he suggested I should take a look round the "meditating ground" and join him later for the main meal of the day.

"About an hour?" I responded, unlocking my legs from their cramped position.

"When you have found what you have found, you can return," was his somewhat mysterious dismissal.

I wandered thoughtfully around the deserted meadow, my gaze constantly lifting to the stunning, final pyramid of Shivling, a banner of streaming ice particles flying from its shining crest...A rash desire surged up inside me. Another fifteen hundred feet before the real climbing started – I could make that, couldn't I? I ran through a mental checklist of my remaining foodstuffs. From the meadow's rim my eye traced the line I would follow... "But the air," a faint voice protested, "won't the air be thinner up there?..and the cold, think of that. And without a tent - !?"

Only when I let all notion of conquest drop from my mind did the tension drain from my body.

Two metal plaques bolted to separate granite boulders caught my attention. They were not of very great age and commemorated two mountaineers who had lost their lives climbing in the region. One was an officer in the Indian army, the other, a woman from Eastern Europe. They had died because, paradoxically, they wanted to live. Did the nature of their deaths demonstrate some valuable human qualities? I like to think so.

Of the mystical outlook, the mountaineer, G.W. Young wrote:

"Through beauty of sight and sensation and through a delight in being alive almost painful in its vividness, there would throb the conviction that all was right with the world and very right with myself and that they were all one and the same thing…"

I stood motionless on the meadow for what seemed like a long while before I heard it. The silence was not that which comes when noise stops. No.

Faint at first, a high, reed-like musical note, a sound so far off, and of such ineffable purity which no earthly instrument could hope to emulate. It quickened, almost to a hum, grew louder like some finely plucked celestial wire: AAAAA…..UUUUUMMMMM…It seemed to have no beginning and no end, stretched irresistibly into the past and future . It welled from the ground, the rocks and the sky, sang both within me and without…No end. I can't describe it further.

How long I remained in this state, I cannot say. Only when I opened my eyes to see a bird flash across my field of vision did I begin to emerge from it.

The sense of heightened reality which I experienced was not simply a reverie or what Tennyson described as 'a kind of waking trance'. I was sure of that. Hindus have numerous paths to this true knowledge, the most familiar being the practice of Yoga, which as understood in India, is very different to its use as a form of body-culture exercises in the West. In Yoga the sound of silence is known as *Anahata Nada*.

Before heading off to his firewatcher's lookout cabin in the Cascade Ranger of the Pacific North West, Jack Karouac avowed, "If I don't have a visionary experience on the top of Desolation Peak, my name's not William Blake".

Things didn't quite work out as he'd hoped, but I think he did right to go alone. I think I did too.

I re-opened my eyes to find that nothing had changed. I blinked and stared about. Everything seemed just as before: *samsara*, the world - there it was. But was it? The colours about me seemed much more vivid than before, blazing forth with a clarity and shimmering intensity which I would never have associated with normal sight. Several puff-ball clouds hung overhead in surreal stasis as if suspended from the sky on invisible threads. The shining mountain ridges appeared to follow the silent meanderings of the half-frozen rivulet; otherwise there was no movement at all. Shivling's summit appeared close at hand and far away simultaneously. I stared up at it fixedly for some time, but without the blind desire of before. I felt there was no necessity for an answer because for that everlasting present, however long it was, I had no more questions.

Gradually, I became aware of ordinary physical reality again, of the cold granite wall pressing through the back of my anorak. I was thirsty. Yet, as I swung about and began to make my way towards my appointment with the *baba*, I realised my appetite had all but disappeared.

Bengali Baba appeared not to have moved in the hour and a half I had been away and was now conversing in Hindi with a young man, perhaps an apprentice hermit, who was listening with absorbed attention. The *baba* waved me back to my sackcloth throne and continued to attend to his new visitor for several minutes longer. I waited patiently until at last their business was concluded, watching with some amusement the undignified four-footed exit. Soon afterwards a large plate of the customary fare was set before me. The *baba* took nothing for himself as I began to eat. When I enquired as to the reason, he answered that he ate his main meal in the evening. After watching me swallow a few mouthfuls in silence, he asked, "Is the food good?"

I made a muffled but agreeable response.

However I soon began to find his close scrutiny of my eating a distraction; I began to eat more rapidly, gulping each mouthful down with scarcely a pause for chewing. It was a fool's plan and before long I began to feel acutely uncomfortable. When sweat broke out on my brow and my breathing became laboured the *baba*, gazing at me with impassive equanimity, enquired mildly, "Are you all right?"

A ringing noise started inside my head as I gasped for water. It had little effect; the discomfiture increased. Throwing all notions of further social graces to the wind I lunged for the door.

Sir Lesley Stephen the eminent Victorian mountaineer, once boasted that he never used his brain above 5000 feet. As I staggered into the freezing air, I wish I had taken the trouble to use mine.

For a few awful seconds I thought I was about to become another newspaper headline in my own country: CHAPATI VICTIM: Tourists Warned. But by degrees the choking sensation diminished on its own accord and five minutes later I was more or less back to normal.

During a few final minutes with Bengali Baba, I asked him if any Westerner had spent a whole winter at Tapovan. I think he sensed the purpose of my question and remained silent for a moment.

"Two years ago," he began, "there was a foreign woman who tried to stay the winter. In the end she had to be carried down, she was so weak."

I indicated my interest with a nod.

"For a long time she drank only ice. That is very bad – there are no minerals in it. Also the altitude was a great problem for this woman. She began to bleed from all parts of her. It is not good for Western people to stay long at this height – you are not as used to it as we are."

More ice was on the way, he told me; therefore it would be in my own interests not to stay the night. He had room but food was low. I took the hint and set off a short while later, lighter by one ten-rupee note which I left behind as a donation.

I had only proceeded a short way from the plateau when I stopped to talk briefly with an older, well-spoken Indian man accompanied by a youth; both were dressed in neat Western-style trekking outfits. I should have known better to ask the question I did.

"No, we are not on a *yatra*," the man answered in peremptory fashion. "We are on a trek!" The last word was so heavily underscored as to make the call to pilgrimage seem like an insult to the Indian nation.

He wasn't the first Hindu I had met who quickly saw fit to distance his worldly self from the 'village gods' aspect of Hinduism. Even a poet born in the Gangotri region but of growing international reputation, had nothing to say about yogis and their kin. Talking with such people was about as fruitful as asking the average optician about inner sight.

I left them and continued in the direction of the flag marking the start of the descent to the glacier. Suddenly, who should pop up from nowhere but the student, Ajay, toiling along on his own. His friends, they were all still sick, he informed me, but he had sufficiently recovered to follow on about two hours after me. The others would await his return.

I didn't detain him any longer since it was already past mid-afternoon and a glacier was no place to be caught wandering during the frozen hours of darkness. Tinting the glacier's pristine whiteness to the north was a corrugation of bluish shadows, a tell-tale sign of concealed crevasses. This time I had the crumbling slope to myself and with gravity's assistance soon found myself once more picking my way between the stony ice waves of the glacier. Some nine hours had elapsed since my leaving Lal Baba's ashram.

Now I had another decision to make. From where I stood I could see several matchstick figures milling about in the ashram compound – another unquiet night was obviously there for the asking. A cramped room with sealed windows bulging with unkempt trekkers, the lost and hungry – it took me only a few moments to decide. Unlike the 'Death Zone' which lies above 28,000 feet, the 'Dread Room', although rarely fatal, may be encountered at any altitude and in any continent. I moved on with a lighter step.

Not far along the higher footpath which contours the Bhojbasa sanctuary however, a sight passed me which I thought had long been consigned to the pages of India's history books.

Two *dandies*, traditional 'sit-up' stretchers used for centuries to carry the aged, rich, the sick and idle to places otherwise inaccessible to them, bumped by on their way to Gaumukh. Four sweating bearers, one to a corner, carried the supine passengers whose aged faces peeped stoically from under thick covering blankets.

I made a brief halt at the *chai* stall. Scarcely had I stretched out my legs when a man rushed up and began to speak urgently to the owner. His expression and body language told me that something had happened.

"A man had been hit on the head by a stone," the *chai wallah* explained.

I needed no telling as to which foreboding spot was probably the scene of the incident. Despite my growing tiredness I felt obliged to offer assistance.

"No, no – help not needed!" came the reply. "He was hit on the head, but he is able to walk."

I subsided into my seat at the cheering news.

Four lean-looking porters in plimsolls and carrying coiled ropes across their shoulders then appeared for a brief stop, perching themselves on the rounded boulders before the shelter.

"Gangotri?" I ventured to the stall-owner when they set off again at a lively pace, still chatting in spirited fashion.

He nodded back, "Gangotri."

I drained my glass and, summoning my reserves of energy, heaved myself to my feet and trudged on after them. I soon topped the short rise which commanded a panoramic view of the return journey.

For a mile ahead the footpath was devoid of life. Where were they? I wondered, unable to see any movement at all on the grey slopes. Only when I extended my gaze did I appreciate how rapidly an unladen porter could move: like beads on a thread, all four were running at full tilt below the danger zone, the very place which awaited me some twenty minutes hence. In the growing twilight of early evening their dancing shapes were almost indistinguishable from the mountainside behind them. Finally, they vanished in quick succession round a bend and I didn't see them again. Once more I had the path to myself. Reassuring though it was to reflect that many thousands had trodden the route before me without harm, I couldn't prevent the arising of a gloomy anecdote some *chai*-shop acquaintance had told me:

"I'd, like, just crossed the place an' was resting on the far side. I heard a noise. I looked back an' the whole slope just poured over the trail an' into the river. Two Indians went with it".

When the girdling footpath brought me into view of the first warning notice, the *chai-wallah's* timely advice to "listen as well as look", rose uppermost in my mind. I forced the pace despite a growing tiredness in my legs, reasoning that the quicker I was beyond that sullen spot the better. Even so, with my rucksack dragging me back, I could not hope to match the speed of the fleet-footed porters. Far ahead, down the deep V-shaped valley, where the evening had already deepened into a darkness in which nothing was distinct or solid, I fancied I could see a friendly twinkling light. Whether a star or lantern I couldn't tell, but there the night sky met the earth and I would be safe.

The angle of the slope increased as I came within range of the stone-scoured gullies. Half-concealed by the fast-fading light, the fan-shaped chutes were loud with portentous silence, while blackened shadows cast amongst the grey only served to heighten the wild and sombre mood. Hereabouts also, the mountain base, the footpath and the fast-running river squeezed close upon each other, which meant I had no alternative but to pass under the very mouths of these boulder-channels. The pinnacles which I described earlier, looked even more menacing than in daylight, leaning over at threatening angles as if keeping enigmatic watch upon the trail.

When a solitary pebble dislodged itself from above, to tinkle harmlessly on to the trail in front of me, my pulsebeat increased and I glanced warily up in case more was to follow. Luckily, there was no further movement and I reached the relative security of the second notice, where I paused briefly to regain my breath. Fifty yards beyond it however, I came upon a sizeable boulder embedded in the footpath.

At this point, although there was no cause for alarm, it became certain that darkness would overtake me well before I reached Gangotri. Happily, there would be a moon to light the way, though not until much later. The next precipitous stretch of the trail, the apron of slabs, was still to come and rather than negotiate that in the dark, I decided to let prudence call the tune and spend the night out. The weather was fine and a few hours sleeping in the open was a better prospect than tramping on, with always the possibility of tripping headlong over an unseen stone. I continued to walk quite quickly, all the while keeping an eye open for a suitable place.

The path descended very gradually and only slowly did the mountainside change its character from one of barren rock to less precipitous slopes supporting an occasional tree or patch of withered brush. My load grew heavier but I remained determined not to stop until I found a spot offering protection from stones which might fall from the heights during the night. Only when the fir-tipped limits of the Chirbasa camping ground appeared did I at last chance upon such a secure bivouac site. A sandy nook by a loop in the trail, it was protected from above by a rock lip, while the trailing arms and roots of a rugged pine offered additional semblance of shelter.

Twelve hours had now passed since I had first left Lal Baba's ashram and I wasted no time in setting out my sleeping bag. This I did after first clearing a lying space of protruding stones and covering it with a light groundsheet. I kicked off my hot boots and settled down full length with a final apple saved from Tapovan. It was too late for anyone else to be on the trail, or so I assumed.

But India is a land which demands constant alertness for the unexpected. And so it proved. I had just begun to relax inside my sleeping bag when I was suddenly alerted by a nearby footfall in the darkness. That I wasn't the only creature wandering abroad at that hour was clear. The footsteps came nearer and stopped.

In such circumstances it is easy to convert an innocent-enough noise into a real-life threat from some predaceous nocturnal animal, and I sat bolt-upright, thrusting my face and arm from the hood of my sleeping bag. Had I been scented by some roaming leopard in search of a meal? No, fortunately. I found myself staring into the mildly pondering features of a stationary mule. At its rear, the eyes of its minder glinted enquiringly at the bundled-up shape confronting them.

"*Namaste,*" a useful password for all occasions relieved the tension: the man relaxed, uttering a quiet word of command. At this, the animal plodded forward a couple of paces, allowing another of its brethren and the first man's companion to emerge from the shadows beneath the tree. The second man took in the situation at a glance.

"*Hotal,* two minutes," he waved his arm to indicate the trees behind him.

If the weather had been bad, then I might well have acted upon the well-intentioned information. But where was the advantage in exchanging a night under the stars for a crowd of restless bedfellows in a repeat of the previous night's entertainment? I shook my head, patting the ground with custodial assurance:

"*Hotal* here. *Thik Hai* (All right). Morning-time Gangotri."

Both men looked down on me in a faintly bemused manner, as if I might not have my wits about me. The second tried once more to convince me of the wisdom of his words, but this time his tone was less self-assured, as though he sensed I would not be easily persuaded. I suppose I might have struck them as being a crazy wanderer from cloudland, though to me my actions seemed sensible enough. Fortunately, the pair gave up at this point, the lead animal was urged forward, and in single-file the procession proceeded on into the tranquil night. When I could no longer hear the mules' hoofs clicking on the rocky trail, I withdrew into my sleeping bag and settled down for the second time.

To my left a skein of milky mist was starting to form across the surface of the silvery Ganga, whilst high overhead remote peaks, hinting at invisible forces, gleamed, illusive, like a dream of a perfect world. Gradually the sky turned a subtle magenta hue; the summits across the river darkened further, save one. For a minute or two its tip was lit by a golden light, then this too faded to a ghostly whiteness. Higher still a nest of fugitive stars glittered like a coronet of heavenly jewels.

In our haste, we sometimes forget the mountains, the stars and the planets of the Milky Way are like ourselves, travellers through time. They, too, follow the Path of Changes. I stared out at the scene for a long time until, finally, feeling my eyelids growing heavier I turned on my right side and went to sleep. Two or three hours later I awoke with stiffness in that hip, so I turned over and lay awake for a short time until I dropped off again. So passed my night out at eleven thousand feet. It had been an eventful day.

I awoke to find the sun already up, a solitary golden ball in an atmosphere of breathtaking clarity. Most of the mist had already dissolved away, but what remained in the valley bottom rolled and bellied like a lake freed from the restraining hand of gravity. Something of the mountain air's cool stillness seemed to have entered me and for a few minutes I sat in silent observance of Nature's timeless wonders.

The peaceful forest dell of Chirbasa was immediately round the corner. With its soft meadow, pines and an unobtrusive stream tinkling between the granite outcrops, it was a perfect place to linger. A homely sign dangling from the door pole of an ancient ten-man ridge tent proclaimed grandly:

<u>WELCOME TO YOUR GOOD NIGHT'S REST</u>

Within the open flap I glimpsed several sleep-crumpled individuals struggling amidst a sea of heaving blankets. This was obviously the '*hotal*', a Klondike relic and best avoided except in the direst of circumstances. I accelerated past it and soon found myself once more upon open ground. Beneath the sky's vaulted arch the shimmering, unblemished snows gleamed a fitting goodbye. Though the sanctuaries of Gaumukh and Tapovan were already shut off from view by the intervening ridges, I had the conviction that their haunting vibration would stay with me for a long, long time to come.

OM NAMAH SIVIA. I sang these words into the flowing, pine-scented stillness, and walked on.

The End

253

APPENDIX
WHAT I TOOK WITH ME

Rucksack, with plastic liner
Small padlock and chain
Sleeping bag and liner
Three T-shirts
Two pairs underpants
One shirt
Pullover
Hankies
Spare pair socks
Flip flops
Anorak
Spare cotton trousers
Long johns
Quilt waistcoat
Scarf
Balaclava
Gloves
Light groundsheet
Space blanket
Towel, toiletries, etc
Small pair scissors
Minimum repair kit
Sun cream and lip salve
Sink plug
Water heater (plug-in type)
Length of twine
Roll of cellotape
Tea infuser
Water bottle
Small torch; candles
Lighter
Knife and spoon
Small cooking pot
Snow goggles

Tube travel wash
Mosquito repellent
Tiger Balm
Garlic Perles (daily)
Sterotabs (water purification)
Small first aid kit incl. Antiseptic
cream, elastoplast etc.
Liv. 52. (protective dose daily)
Himalayan Drug Co.
Fleaseed Husk (*Sat-Isabgol*) Stomach
disorders, diarrhea
Rehydration mix
Camera, spare film
Maps, guidebook notes, notebook
Photo-copies of documents
Dried soups
Tea bags
Glucose
Dried fruits
Prana balls, (fresh)
Dried milk
Two books to read; Hindi phrase book
Patience (lots)

INDEX

Cover Design: The Digital Canvas Company
                  Forres
                  Scotland
                  bookcovers@digican.co.uk

Layout: Stephen M.L. Young
                  Elgin
                  Scotland
                  stephenmlyoung@aol.com

Font: Adobe Garamond (11.5pt)

Copies of this book can be ordered via the Internet:

                  www.librario.com

or from:

                  Librario Publishing Ltd
                  Brough House
                  Milton Brodie
                  Kinloss
                  Moray IV36 2UA
                  Tel /Fax No 01343 850 617